VI International Scientific Conference of Young Scientists "Applied Mechanics and Dynamics Systems"

Devoted to the 75th Anniversary of the National Research Nuclear University MEPhI and the 95th Anniversary of Academician N.G. Basov

Journal of Physics: Conference Series Volume 945

Moscow, Russia
17 – 21 April 2017

ISBN: 978-1-5108-5752-0
ISSN: 1742-6588

Printed from e-media with permission by:

Curran Associates, Inc.
57 Morehouse Lane
Red Hook, NY 12571

Printed by Curran Associates, Inc. (2018)

For permission requests, please contact the Institute of Physics
at the address below.

Institute of Physics
Dirac House, Temple Back
Bristol BS1 6BE UK

Phone: 44 1 17 929 7481
Fax: 44 1 17 920 0979

techtracking@iop.org

Additional copies of this publication are available from:

Curran Associates, Inc.
57 Morehouse Lane
Red Hook, NY 12571 USA
Phone: 845-758-0400
Fax: 845-758-2633
Email: curran@proceedings.com
Web: www.proceedings.com

Table of contents

Volume 945

VI International Scientific School-Conference of Young Scientists "Modern problems of physics and technologies", devoted to the 75th Anniversary of the National Research Nuclear University MEPhI and the 95th Anniversary of Academician N.G. Basov

17–21 April 2017, Moscow, Russian Federation

Accepted papers received: 18 December 2017
Published online: 31 January 2018

Preface

Papers

VI International youth scientific school-conference «Modern problems of physics and technology», dedicated to the 75th anniversary of the MEPhI and the 95th anniversary of Academician N.G. Basov

O N Krokhin[1,2], I N Zavestovskaya[1,2], I I Yashin[1], A A Ananskaya[1,2] and A A Fronya[1,2]

[1] National Research Nuclear University "MEPhI", Moscow, Russia
[2] P.N. Lebedev Physical Institute of the Russian Academy of Sciences, Moscow, Russia

The annual International Youth Scientific School-Conference "Modern problems of physics and technology' has become a traditional spring event, organized in MEPhI. In 2011, the School-Conference was conceived as a scientific and practical school for students, graduate students and young scientists, where leading scientists represented scientific and educational lectures on modern and advanced research areas in science and technology. The first Youth Scientific School was organized by the Faculty of MEPhI N.G. Basov Higher School of Physics, and was dedicated to the 40th anniversary of the Faculty, attracted the attention of more than 100 participants from various regions of Russia and CIS countries.

At all times, educations agree and meet the requirements of modern society. The main task of modern education is not only the training of specialists necessary for society, but also the upbringing of a creatively active person who is capable of independent activity, striving for further self-improvement, with a high level of motivation for learning. Modern trends in the organization of the educational process are aimed at the development of creative thinking and the disclosure of creative potential, which is realized through the early involvement of the student in research activities. For revealing creative youth, stimulating interest in science, technology, engineering activities, there is a need to organize such events as scientific and practical schools, competitions, olympiads, etc. The International Youth Scientific School-Conference "Modern Problems of Physics and Technologies" serves to solve such problems. The main goal of the School-Conference is attracting young active cadres to science and education. In addition School-Conference pursues other important goals, such as creating favorable conditions for self-expression and self-realization of talented youth, stimulating scientific and technical creativity, identifying the actual needs of science and industry in training, forecasting of prospective professional fields. Currently the revealing, development and support of talented youth in Russia can be seen as a concerning for future science, technology in Russia, and Russia's position on the world stage.

At present, the School-Conference is a platform that combines:

- Lectures of leading foreign and Russian scientists on modern promising research directions
- Speech by students, graduate students, young scientists with oral and poster presentations
- Participation of students and young scientists as school listeners without presentation of the report

• Competition of young scientists, post-graduate students and students reports with the possibility of publishing articles in the journals Web of science, Scopus

• Competition of school project works

• Organization of sightseeing tours of modern laboratories and scientific centers of MEPhI, as well as partners.

The VI International Youth Scientific School-Conference "Modern Problems of Physics and Technologies" was held at the MEPhI from 17 to 22 April, 2017. The conference was organized by the N.G. Basov Higher School of Physics and the Institute of Engineering Physics for Biomedicine MEPhI, as well as the P.N. Lebedev Physical Institute of the Russian Academy of Sciences. This year the School-Conference is devoted to two significant events - the 75th anniversary of the MEPhI and the 95th anniversary of the academician of the Russian Academy of Sciences, the Nobel laureate, the graduate of MEPhI, the founder of the Higher School of Physics (earlier the Special Physics Faculty), Nikolay Gennadievich Basov. The anniversary program of the School-conference included reports and lectures on the advanced research directions conducted at the MEPhI, reports of students and followers of N.G. Basov, and graduates of the N.G. Basov Higher School of Physics.

From year to year the scale of the School-conference is increasing, the geography of participants is expanding, the level of the presented works is increasing. In 2017, 262 delegates from different parts of our country, CIS countries and foreign countries took part in the School-Conference. 13 countries and more than 40 cities of Russia were represented.

The program of the School-conference provided 4 plenary sessions, 2 poster sessions and 3 oral youth sessions, at which participants had the opportunity to listen to the lectures of the world's leading scientists and make reports on the results of their own work on the main topics of the School-conference:

• "Laser physics, solid state physics, optics and plasma physics"

• "Engineering Physics for Biomedicine"

• "Nuclear physics and technology, particle physics, astrophysics and cosmophysics"

• "IT - technologies, intellectual systems, cybersecurity"

• "Electronics, automatics, spintronics, nanotechnology".

Lecturers of the school-conference are people with a worldwide reputation:

Academicians of Russian Academy of Sciences Oleg N. Krokhin (P.N. Lebedev Physical Institute, MEPhI, Russia), Boris Yu. Sharkov (FRRC-FAIR Russia Research Centre, Germany, MEPhI, Russia), Yurii N. Kulchin (IACP, Far-East Division of the Russian Academy of Sciences, MEPhI, Russia); Yuris Purans (University of Latvia, MEPhI, Russia), Igor V. Meglinski (University of Oulu, Finland, MEPhI, Russia), Giovanni Barbero (Politecnico di Torino, Italy, MEPhI, Russia), Andrey Kabashin (Aix-Marseille University, France, MEPhI, Russia), Maxim M. Nazarov (NRC "Kurchatov Institute", Russia), Vladimir Lysenko (National Institute of Applied Sciences, Lyon, France, MEPhI, Russia), Rudolf Wolfgang Steiner ("Institute for Laser Technologies in Medicine and Metrology (ILM)" at the University of Ulm, Germany, MEPhI, Russia), Alexander V. Parfenov (Physical Optics Corporation, USA), Anton Fojtik (Czech Technical University in Prague, Czech Republic, MEPhI, Russia) and others.

Annually in the framework of the School-conference a competition of reports of young scientists, post-graduate students and students is held. So in 2017, more than 200 oral and poster reports were heard and reviewed. According to the results of the School-Conference in 2017, 3 winners were selected in each section (one among the oral presentations and two among the poster sessions), 12 papers were recommended for publication in the journal "Bulletin of the Lebedev Physics Institute", and 34 papers for publication in the Journal IOP: Conference Series. The most original report was also selected. On the results of the competition the winners were awarded personal certificates for the right to publish an original scientific article.

This year a pleasant innovation for the School-conference was the Competition of School project works, in which 32 schoolchildren together with their scientific adviser took part. The poster session for schoolchildren was held on April 20 in a very cozy and friendly atmosphere. The pupils presented very interesting works, over which they worked more than a year. Among the presented works were the winners of the “Junior” competition. The Competition Commission appreciated all the presented projects and chose the best ones. Separately, it should be noted the absolute winner of the Competition, which scored the maximum possible number of points - Golovan Olga Daniylovna with her work “Beyond the possibilities. The power of thought”. The purpose of work was to create a neuro-computer interface designed for the study of human biorhythms. Winner is a schoolgirl of the 11th form of the Gymnasium of the Fryazino city.

The VI International Youth Scientific School-Conference once again brought together scientists with world renowned students, graduate students, young researchers, and this year also schoolchildren at the MEPhI. Thus School-Conference is providing all participants with a unique opportunity to exchange experiences and information with their colleagues, the opportunity to make new contacts in the scientific environment, find like-minded people, future scientific leaders, and possibly a future workplace. This is good practices, which necessarily continue. The School-conference is transformed with modern trends in science and education and promotes the revealing of talented youth, and with the involvement of schoolchildren this year as participants in the conference, this event promises to become even more popular and interesting.

Acknowledgments
To express deep gratitude to the lecturers for the reports presented at the VI International Youth Scientific School-Conference.

To express gratitude to the organizers of the School-Conference for conducting the School-conference at a high scientific level.

To express gratitude to the sponsors: MEPhI, P.N. Lebedev Physical Institute.

IOP Conf. Series: Journal of Physics: Conf. Series **945** (2017) 011002 doi:10.1088/1742-6596/945/1/011002

Peer review statement

All papers published in this volume of *Journal of Physics: Conference Series* have been peer reviewed through processes administered by the proceedings Editors. Reviews were conducted by expert referees to the professional and scientific standards expected of a proceedings journal published by IOP Publishing.

IOP Conf. Series: Journal of Physics: Conf. Series **945** (2017) 012001 doi:10.1088/1742-6596/945/1/012001

Porous silicon for drug delivery systems

E N Abramova, A M Khort, A G Yakovenko, D S Kornilova, E A Slipchenko, D I Prokhorov, V I Shvets

Department of material science and technology of functional materials and structures, Moscow technological university, University of fine chemical technologies, prospect Vernadscogo, 86, Russia

anavenko@yandex.ru

Abstract: The article deals with main principles of the formation of porous silicon (por-Si) to produce containers for drug delivery systems. Most important por-Si characteristics to produce nanocontainers with required parameters are determined.

1. Introduction

Using nanostructured containers fordirectional drug delivery is one of the fast developing trends in the medical field. Firstly, it is connected with the low effectiveness of traditional ways of introducing drugs and consequent negative effects. So, the development of undangerous container's materials for the aims mentioned above is one of the most important tasks for medicine and material science. Porous silicon (por-Si) is extremely promising material for such containers. In the first turn it is stipulated by its bio-compatibility, bioaccessibility, biodegradation and nontoxicity. Despite being complicated, electrochemical processes of por-*Si* formation are rather technically simple to carry out and inexpensive [1].

Main characteristics of the material for its use as containers are the radial and axial pore dimensions, pore partition width, porosity, surface extension measure. These parameters depend on the silicon wafer properties (the conductivity type, dopant concentration, structure imperfections, etc.) as well as the porous etching conditions (electrolyte solutions composition, current density, etching time, etc.).

To take into consideration all the factors mentioned above, which are very important to obtain por-Si with previously assigned parameters, is extremely difficult and likely unachievable.

That's why in our opinion to get por-Si with required parameters it is more reasonable to determine the general principles of por-Si formation and how porous layers parameters are dependent on them. Such an approach will let choose specific etching conditions to obtain materials with characteristics, matched the needed extents.

Solution for the problems is the main aim of the research.

2. Experimental

We used n- and p-type silicon wafers with the resistivity in the range $\rho = 0.01 - 100$ ohm·cm, oriented long the (100) and (111) crystallographic planes. The wafers were etched with hydrofluoric acid solutions (45% HF) in H_2O in the volume ratio 2:1 in a galvanostatic mode. The etching cell, made of Teflon, allowed for one-sided horizontal etching. The anode used was a polished massive copper plate, and the cathode had the form of a thin platinum plate. The electric field applied during etching was varied by diaphragms of various sizes and changing the electrode separation.

The nanopore size, shape, and orientation ant the thickness of NS layers were determined using a JEOL electron microscope and a POLARM R optical microscope.

3. Experimental results and discussion

Historically the formation of por-Si was considered as interaction between silicon and F^- ion in HF solutions [1-4]. The correlation between the etching process parameters and the parameters of por-Si was explained on the basis of this approach.

We have revealed [5-7] that porous etching of Si is stipulated not F^- ion, but $(HF_2)^-$ ion, which is contained in HF solutions > 1mol/l.

Difference between the two ions, including their sizes, structures, anisotropy, causes inequality in the interaction of Si with F^- and Si with $(HF_2)^-$, what was described in our articles [4]. Thus, it influences the pore formation in Si with various parameters.

To produce containers of por-Si with required parameters one should process silicon wafer to get porous material with appropriate characteristics. From this point of view, the most important por-Si ones are radial and axial pore dimensions, pore partition width, porosity, surface extension measure. Therefore, this article focuses on general regularities determining the characteristics mentioned above.

On the first figure (figure1) there is the influence of etching time (t) on the thickness of porous layers obtained on n- and p-type silicon wafers with various dopant concentration under various current density (j). The figure 1 shows, that under processing conditions being equal, the thickness of n-type porous layers significantly overtops the thickness of p-type porous layers. The thickness of por-Si layers, obtained on silicon wafers with P dopant (n-type) $8\cdot10^{14}$ cm^{-3} (ρ= 5 ohm·cm), $4\cdot10^{14}$ cm^{-3} (ρ= 10 ohm·cm), j=25 mA/cm^2, t=30 minutes, was 180-220 micron. The thickness of por-Si layers, obtained on silicon wafers with B dopant (p-type) $1.2\cdot10^{15}$ cm^{-3} (ρ= 10 ohm·cm) with other conditions being equal, did not exceed 50-60 micron. Current density from 1 mA/cm^2 to 50 mA/cm^2 does not result in essential thickness changes under the same etching time.

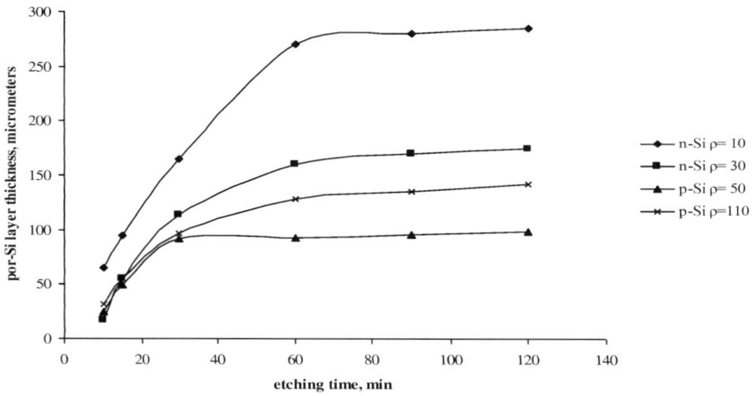

Figure 1. The dependence of por-Si layers thickness on the etching time.

According to the results [8] pore axial dimension depends on two simultaneous processes: Si surface etching with F^- ions (some quality of which constantly exist in HF solutions of any concentration) and porous etching with $(HF_2)^-$ ions. It means that axial pore dimension is limited by the equilibrium between rates of the two processes. This fact should be taken into account to predict axial pore dimension (h) (equation 1).

The figure 2 shows the dependence between pore frequency function and electric-field intensity. The pore frequency function increases while the electric-field intensity decreases.

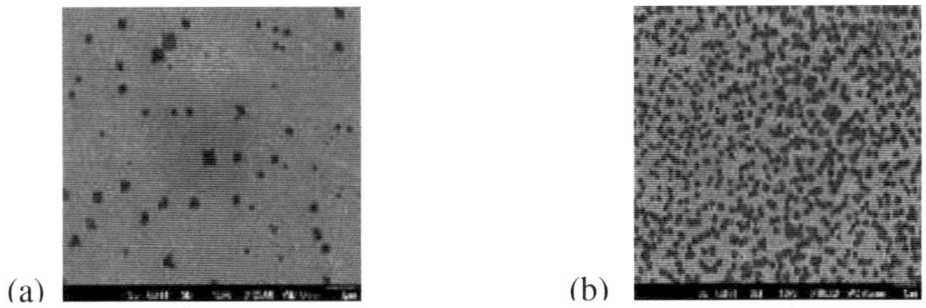

Figure 2. The dependence of pore frequency function on electric-field intensity: the distance between the cathode and the anode (a) 2,5 cm, (b) 7 cm

The influence of the dopant concentration on the pore frequency function can be esteemed on the figure 3, which shows how pore radial dimension depends on the dopant concentration.

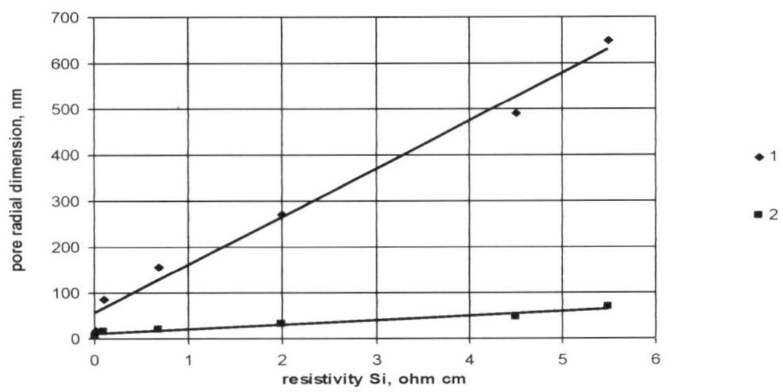

Figure 3. The dependence of the pore radial dimension on Si wafer resistivity: (1) n-Si, (2) p-Si.

According to [8-9] parameters of por-Si more suitable for producing drug delivery systems are 5-25 nm pore radial dimensions and 50% porosity.

From figure3 one can see, that the near-linear correlation between pore radial dimensions (10-15 nm, ρ=0.01 ohm/cm; 600-700 nm, ρ=10 ohm/cm) and Si resistivity (ρ) are observed over the 0.01 to 10 ohm/cm range. The dopant type does not reveal any significant influence [2].

Pore partition widthes in por-Si, obtained on wafers with ρ>0.1 ohm·cm, have the size near to the pore radial dimensions and much more than the pore radial dimensions for ρ<0.1 ohm·cm.

The quality of medical agent in por-Si includes the drug adsorbed in pores as well as on the surface of porous silicon. This fact is very important for medicines with high adsorbing capacity. From these point of view research results of the dependence of por-Si formation from silicon wafer crystallographic orientation is of great importance.

Therefore, investigations of the principles of pore orientation on the wafer are very important.

On the figure 4 there are the photos of por-Si layers, made on (100) and (111) wafers. Pores, generated on (100) wafers, are perpendicular to the substrate. As to pores on (111) wafers, the processes of their generation and growth are more complicated.

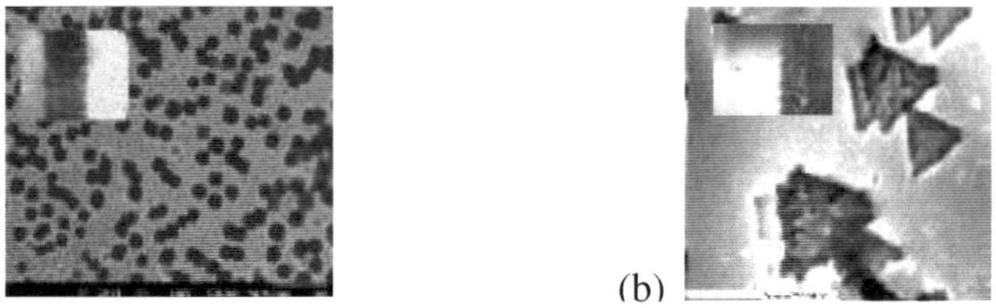

Figure 4. The influence of wafer crystallographic orientation on the pore formation mechanism: (a) Si-wafer (100), (b) Si-wafer (111).

Firstly tetrahedral regions form. Then semi-squared pores grow from the tetrahedral region walls. So that, they are at some angle the substance surface. Such a mechanism of por-Si formation results in branched structures and large internal surface. It let these structures absorb more medicine.

While planning to make por-Si with necessary characteristics, it is beneficial to esteem the influence of the processing parameters. It can be done with equation (1):

$$S = A\frac{m_1}{phN} \tag{1}$$

S – pore radial dimension, cm^2; A – constant, stipulated the mechanism of Si and $(HF_2)^-$ interaction (we have calculated it to be 0.0625); m_l – mass of Si removed during porous etching, g; ρ – the density of Si, g/cm^3 ; h – porous layer thickness, cm; N –the number of pore initiation places per unit area. N depends on dopant concentration C $N = f(C^{\frac{2}{3}})$ under the given electric-field intensity. The data obtained let predict pore sizes to some extent. Data, calculated on the basis of equation 1, satisfactorily coincide.

4. Conclusions: main principles of por-Si formation are determined. It let choose one single silicon wafers and etching parameters to obtain por-Si with parameters required for drug delivery systems on its base and facilitate the choice.

References

[1] Memming R, Sckwandt G Anodi 1966 *J. Surface Scien.* **4** 109-24
[2] Lehman V, Stengl R, Luigart A 2000 *J. Materials Science and Engineering* B **69-70** 11-22
[3] Bisi O, Ossicini S, Pavesi L. 2000 *J. Surface Science Report* **38** 1-126
[4] Goryachev D N, Belyakov L V, Sreselli O M 2004 *J. FTP* **38** 735-9
[5] Abramova E N, Gvelesiani A A, Khort A M, Yakovenko A G 2014 *Russian Journal of Inorganic Chemistry* **59** 1574-8
[6] Abramova E N, Khort A M, Yakovenko A G, Shvets V I 2015 *J. Neorganicheskie marterialy* **51** 815-22
[7] Abramova E N, Khort A M, Yakovenko A G, Shvets V I 2015 *J. Doklady chemistry* **462** 48-51
[8] Ishenko A A, Fetisov G V, Aslanov P A 2011 *Nanosilicon: properties, formation, appliance, methods of use and control* (Moscow: Phismatlit) p 648
[9] Gavrilov S A, Belov A N 2014 *Electrochemical processes in the technology of micro- and nanoelectronics* (Moscow: RIOR, INFA-M) p 239

Raman diagnostics of silicon nanocrystals dissolution in aqueous medium

A F Alykova[1], I N Zavestovskaya[1,2], V G Yakunin[3], V Yu Timoshenko [1,2,3]

[1]National Research Nuclear University "MEPhI", PhysBio Institute, BioNanoPhotonics Lab., 115409 Moscow, Russia
[2]Lebedev Physical Institute of RAS, 119333 Moscow, Russia
[3]Lomonosov Moscow State University, Physics Department, 119991 Moscow, Russia

E-mail: waiste15@bk.ru, vtimoshe@gmail.com

Abstract: Raman scattering in ensembles of silicon (Si) nanocrystals (NCs), which were prepared by differed methods and then stored in an aqueous medium, was studied to reveal features of the Si NC dissolution process. Arrays of crystalline Si nanowires, mesoporous Si nanoparticles and spherical Si NCs have been analyzed after exposure in water during dialysis for elimination of the dissolution products from the medium. According to the one-phonon Raman spectroscopy the Si nanowires with cross-section size of the order of 100 nm were resistant to the dissolution process both in water and physiological saline for 2 weeks at least. The storage of mesoporous Si nanoparticles in water for 1 day resulted in a strong decrease of the Raman intensity and a low energy shift of the Raman spectrum, which was explained by size reduction of Si NCs below 10 nm. Si NCs with the initial size distribution from 5 to 100 nm after exposure in water during 24 hours revealed a decrease in the contribution of small NCs that narrowed the size distribution. The obtained results can be used in order to choose appropriate type of Si NPs for biomedical applications.

Keywords: silicon nanocrystals, nanowires, porous silicon, Raman scattering, biodegradation

1. Introduction

In recent years, various types of silicon (Si) nanocrystals (NCs) and nanoparticles (NPs) have been extensively studied for applications in biomedicine, including both the diagnostics and therapy (theranostics) of cancer [1, 2]. Laser-ablated Si NPs [3] and porous ones formed by electrochemical method [4-8] exhibit unique physical and chemical properties, which can be used for the theranostics of cancer [2, 4, 6, 7]. For example, Si NPs can act as sensitizers for sonodynamic [6, 7], photodynamic [8, 9] and photothermal [10] therapy modalities as well as for the radio-frequency electromagnetic hyperthermia [11] of cancer both *in vitro* and *in vivo*. It is found that Si NPs reveal properties of contrast agents for optical [3-5] and magnetic resonance [12] bioimaging. Photoluminescence (PL) of laser-ablated Si NCs with sizes of 3-5 nm was shown to be efficient in cancer cells *in vitro* [3]. Bioimaging studies with detection of the PL and Raman signals from porous Si NPs were used to monitor processes of the uptake and biodegradation (dissolution) of NPs in cancer cells *in vitro* [5].

In our work Si NPs, which were prepared by different chemical and physical methods, were stored in aqueous media and then they were investigated by means of the Raman spectroscopy in order to reveal features of the dissolution process.

2. Samples and Methods

Experiments were carried out with samples of Si nanowires (NWs) grown on crystalline Si (c-Si) substrates, mesoporous Si (meso-PSi) NPs and non-porous Si NCs. Figure 1 shows typical images of the samples in a transmission electron microscope (TEM).

Si NWs were formed by using metal-assisted chemical etching (MACE) of boron doped (100)-oriented c-Si wafers with specific resistivity of 1-10 Ω·cm. The MACE process has been done in two steps [13]. At the first step, silver (Ag) NPs with mean sizes 20-100 nm were deposited on the top of c-Si substrates by using the chemical reduction of silver nitrate ($AgNO_3$) in a solution of hydrofluoric acid (HF) [14]. Afterwards, the c-Si substrates with deposited Ag NPs were etched in a mixture of HF and hydrogen peroxide (20%) in water that resulted in heterogeneous Ag NP-assisted etching of c-Si, which was transformed into Si NW arrays [13, 14]. Then residual Ag NPs were removed by etching in concentrated nitric acid and the Si NWs on c-Si substrates were carefully washed in de-ionized water.

Figure 1a shows a TEM image of the prepared Si NWs detached from the substrate and deposited on a carbon grid. One can see from the figure that the NWs have cross-sectional size (diameters) about 20-100 nm. The length of SiNW was controlled by the time of the second step of MACE and it was about 10-15 μm for the investigated samples. The prepared NWs on c-Si substrates were immersed in de-ionized water or in saline (aqueous 0.9% NaCl solution). Prior to Raman measurements the samples were removed from the solution and they were dried in air at room temperature for 20 min.

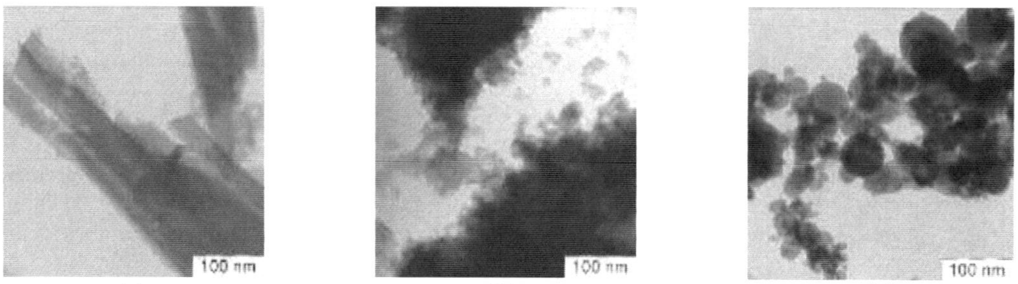

(a) (b) (c)

Figure 1. TEM images of (a) Si NWs, (b) meso-PSi NPs and (c) Si NCs.

Samples of meso-PSi were prepared by standard electrochemical etching of heavily boron-doped c-Si wafers with surface orientation (100) and specific resistivity of 1-5 mΩ·cm in a solution of HF and ethanol (HF(50%):C_2H_5OH=1:1) at etching current density of 60 mA/cm^2 for 1 h (see, for example [4]). The prepared meso-PSi films with thickness about 70 μm were separated from c-Si substrates by a short increase of the current density up to 600 mA/cm^2. Aqueous suspensions of meso-PSi NPs were obtained by mechanical grinding of meso-PSi films mixed with de-ionized water. The grinding was done by using a planetary ball mill FRITSCH "Pulverisette 7" with zirconium oxide mill vessel and balls (diameter of 5 mm) at rotation speed of 800 rpm for 9 minutes. According to the TEM data (Figure 1b) as-prepared meso-PSi NPs are characterized by mean size of about 50-100 nm in agreement with previously reported results [4]. It is seen from Figure 1b that the meso-PSi NPs have porous structure and they consist of agglomerates of Si nanocrystals with smaller sizes.

Plasma-synthesized Si (pl-Si) non-porous NPs were prepared by microwave plasma-assisted ablation/fragmentation of microcrystalline Si powder (99.9%) in inert gas atmosphere (see for details Ch. 5 in [1]). The TEM study shows that initial pl-Si NPs consist of nearly spherical Si NCs with sizes in the range from 5 to 100 nm (see Figure 1c).

The dissolution of NPs of meso-PSi and pl-Si in aqueous medium was studied for the samples with initial concentration about 1 g/L, which were put in dialysis bags with pore sizes of 6-10 kD and kept for 1 to 24 hours in distilled water at room temperature. Then, at intervals 1, 2, 3, 6, 12, 24 hours, drops of the suspension were taken and deposited on stainless steel substrates and then they were dried in air for 10-15 min.

Raman spectra of the dried samples were measured by using a double gratings DFS-52 spectrometer (LOMO, Russia) in a 90-degree scattering geometry. A cooled photomultiplier FEU-79 served as the detector. The source of excitation radiation was an argon laser with wavelengths of 488 and 514.5 nm and excitation intensity on the sample was limited to 100 W/cm^2 to avoid the laser-induced heating To measure the Raman spectra with excitation at the wavelength of 632.8 nm, a micro-Raman spectrometer Horiba Jobin Yvon LabRAM HR Visible was used.

3. Results and Discussion

Figure 2 shows the Raman spectra of the initial sample of SiNWs and those stored in water or in saline. The spectra consist of a sharp line at 520.5 cm^{-1} with FWHM (full width at half maximum) of about 3-4 cm^{-1}, which corresponds to the scattering by optical phonons of the crystal lattice of Si. The high intensity line in comparison with c-Si substrate indicates an enhanced interaction of the exciting radiation with SiNWs due to the multiple scattering. The line shape of the Raman spectra of the both c-Si and SiNWs is well described by the Lorenz function:

$$I_C(\omega)=\frac{C}{(\omega-\omega_c)^2+(\frac{1}{2}\Delta\omega_c)^2}, \qquad (1)$$

where ω_c = 520.5 cm^{-1} is the Raman phonon frequency in c-Si, $\Delta\omega_c = 3\,\text{cm}^{-1}$ is the width of the Raman line of c-Si at room temperature, C is a constant.

It was found that exposure of SiNWs in aqueous solutions for 2 weeks at least did not lead to a significant change in line shape (see Figure 2), indicating the immutability of the dimensions of nanowires. The reason for this fact can be the significant size of the cross section of SiNWs and sufficient quality oxide layer on the surface of SiNWs formed at the final preparation stage.

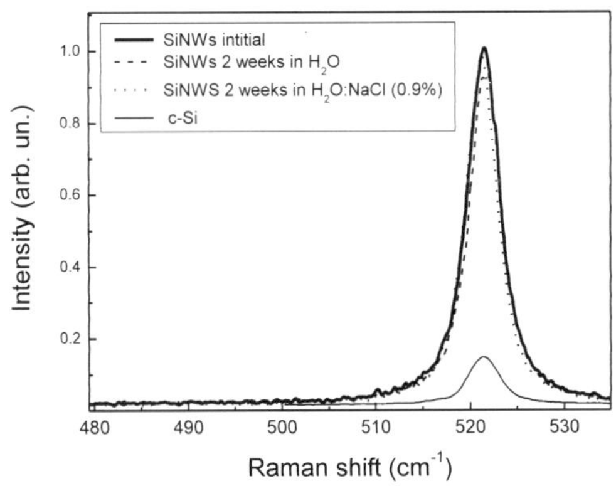

Figure 2. Raman spectra of SiNWs before (thick solid line) and after storage in water (dashed line) and saline (dotted line) for 2 weeks as well as for c-Si substrate (thin solid line), for comparison.

Figure 3 shows Raman spectra of meso-PSi NPs before (as-prepared) and after storage in water (in dialysis regime). Note that the Raman scattering intensities for these samples were rather low relative to the signal from undoped c-Si wafers that can be explained by the Fano effect due to a significant number of equilibrium charge carries (holes) in meso-PSi [15]. The meso-PSi NPs before storage in water exhibit an asymmetric line with maximum at 518 cm^{-1}, full width half maximum about 8-9 cm^{-1} and a shoulder in the low frequency part of the spectrum. Such form of the Raman spectrum indicates an effect of the phonon confinement in Si NCs with sizes about 5 nm [16]. Beside the NC line the spectrum consists of a week and broad band with maximum at 480 cm^{-1}, which is typical for amorphous silicon (a-Si) [15]. Indeed, a certain part of meso-PSi NPs can be amorphous due to the procedure of mechanical grinding in water, which was used for the NP preparation.

To fit the Raman spectra of meso-PSi we use a model proposed in Ref.[17]. According to the model the Raman spectrum of Si NCs is described by the following equation [16,17]:

$$I_{NC}(\omega)=\int_0^1 \frac{B\cdot\exp\left(-\frac{1}{4}q^2L^2\right)\cdot q^2}{(\omega-\omega_c(q))^2+\left(\frac{1}{2}\Delta\omega_c\right)^2}dq, \quad (2)$$

where ω_c = 520.5 cm^{-1} is the Raman phonon frequency in c-Si, parameter $L = d/a_0$, d is the diameter of NC, a_0 = 0.543 nm is the lattice constant of *c*_Si, q is the phonon wave vector expressed in $2\pi/a_0$ units, and $\omega_c(q)= \omega_c(1 - 0.18q^2)$ is the phonon dispersion relation in c-Si [17], B is a constant. Note, that Equation 2 well describes the phonon confinement in spherical Si NCs with sizes 1-10 nm [16], while for the Raman scattering from larger NC one can simply use Equation (1).

The Raman spectrum of a-Si is described by the following relations [17]:

$$I_A(\omega)=A\cdot\exp\left(-\frac{(\omega-\omega_a)^2}{2\delta^2}\right), \quad (3)$$

where A is a constant, ω_a = 480 cm^{-1} and $\delta=\Gamma\left(2\sqrt{2\ln 2}\right)^{-1}$, Γ = 70 cm^{-1} are the position of the peak and its width at half maximum, respectively.

It is assuming that the Raman scattering from the investigated sample consists of contributions of all fractions (phases), i.e. NCs with size above 10 nm without phonon confinement (c-Si-like fraction), small Si NCs with phonon confinement and a-Si fraction, and it can be given by the following expression:

$$I_{Sum}(\omega)=I_C(\omega)+I_{NC}(\omega)+I_A(\omega)\ . \quad (4)$$

The volume fraction of each phase can be estimated as a ratio of the corresponding integrated intensity to the total Raman spectrum, e.g. $f_{NC}=\int I_{NC}(\omega)d\omega \Big/ \int I_{Sum}(\omega)d\omega$.

Fitting of the Raman spectra of meso-PSi by using Equations (1)-(4) shows that the mean size of Si NCs accounts 5 nm and it remains nearly the same for as-prepared NPs and stored in water ones, while contribution of the Si NCs becomes smaller and the a-Si fraction increases (see Figure 3). The latter indicates additional disordering of the crystal lattice of Si NCs due to their partial dissolution in water in agreement with [5].

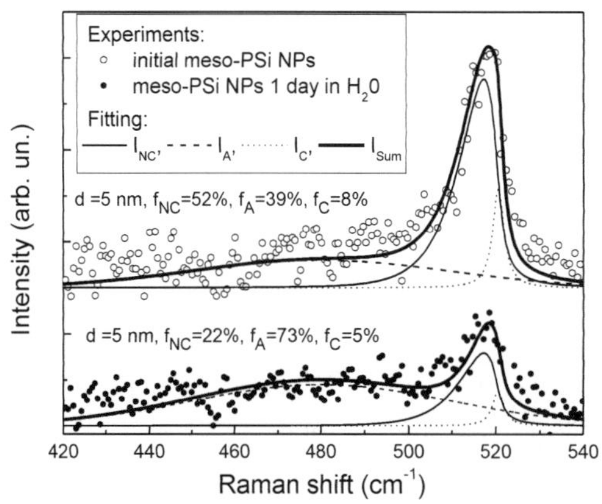

Figure 3. Raman spectra of as-prepared meso-PSi NPs (open circles) and those after 1 day storage in water (filled circles) and fits (lines) of the corresponding spectra by using Equations (1)-(4).

Figure 4 shows Raman spectra of pl-Si samples before and after storage in water (dialysis). The spectrum of the initial sample is asymmetric with maximum at 520 cm^{-1} and a width of 8 cm^{-1} that indicates the presence of Si NCs with a wide size distribution from 5 till 100 nm in agreement with the TEM data (see Figure 1c). Upon exposure of the samples in water for 1 hour a significant decrease in the intensity and change of the spectrum of Raman scattering occurred, which can be explained by dissolution of part of the NCs and the change in their size distribution. The sample stored in water for 24 h exhibited a strong decrease in the intensity of Raman scattering, and the position (520 cm^{-1}) was similar to that for c-Si substrate. This fact indicates almost complete dissolution of small NCs with sizes below 10 nm, while NSs with large sizes remained. Indeed, in the TEM image of pl-Si show NPs with sizes of about 100 nm (see Figure 1c), which do not subject to significant reduction of their size during a one day in water. On the one hand the latter fact is similar to the stability of SiNWs (see Figure 1a). On the other hand, the crystal lattice defects and/or large specific surface area, which are inherent for small pl-Si NPs, are possible reasons for their rapid dissolution in water as it was observed in the Raman scattering measurements (see Figure 4).

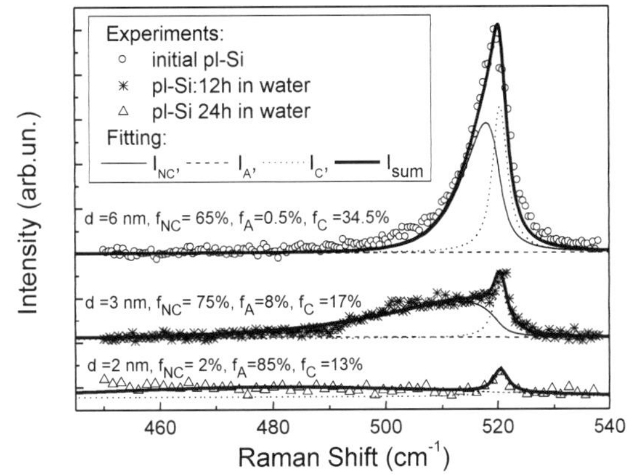

Figure 4. Raman spectra of pl-Si samples before (open circles) and after storage n water for 12 h (stars) and 24 h (up-triangles). The spectra are fitted by Equations (1)-(4) (I_{Sum}, thick solid lines), while the fractions of NCs (I_{NC}), amorphous (I_A) and crystalline (I_C) phases of Si are plotted by thin solid, dashed and dotted lines, correspondingly. The curves are translated in the vertical direction for better representation.

4. Conclusions

The Raman spectroscopy studies of three types of nanocrystalline silicon samples with different nanocrystal sizes and morphologies, i.e. crystalline nanowires, mesoporous nanoparticles and spherical non-porous ones, were carried out to reveal features of their dissolution in aqueous media. The nanowires formed by metal-stimulated etching of c-Si, were found to be resistant to the dissolution both in pure water and in saline solution for 2 weeks at least. It can be explained by the crystal structure, relatively large cross-sectional size about 100 nm and the oxide-coated surface quality. In the series of experiments with meso-PSi NPs, it was found that dissolution of the significant part of nanocrystals occurred after the storage in water for 1 day, was accompanied by the increase of the relative contribution of the amorphous silicon phase. As for the non-porous nanoparticles it seems that the nanocrystals with sizes less than 10 nm are almost dissolved within 1 day of storage in water, while nanocrystals with sizes ~100 nm are retained. The obtained results allow us to choose an appropriate type of Si nanostructures for biomedical purposes. In particular, silicon nanowires, because of their stability in aqueous medium, can be recommended for use as active elements of biosensors or biomarkers in biochemical sensors. Nanoparticles of mesoporous Si as quickly biodegradable material can be used for express therapy of various diseases, e.g. hyperthermia of cancer. As for the non-porous nearly spherical Si nanoparticles, which exhibit the strong Raman scattering even after one-day storage in water, can be considered as an agent for the simultaneous optical diagnostics and therapy (theranostics).

Acknowledgments
The authors thank Prof. A. A. Ishchenko for providing with nanocrystalline silicon samples, Dr. A. V. Pavlikov for the assistance with micro-Raman measurements, Dr. S. A. Abramchuk for the TEM measurements and Dr. A. Yu. Kharin for useful discussions. Prof. V.Yu.Timoshenko acknowledges the financial support by the Ministry of Education and Science of Russia (project No.16.2969.2017/4.6 at the PhysBio Institute of National Research Nuclear University "MEPhI").

References

[1] Ischenko A A, Fetisov G V, Aslanov L A, 2014 Nanosilicon: Properties, Synthesis, Applications, Methods of Analysis and Control, *CRC Press* **2** 660.
[2] Kabashin A V, Timoshenko V Yu 2016. What theranostic applications could ultrapure laser-synthesized Si nanoparticles have in cancer? *Nanomedicine* **11** (17) 2247-2250.
[3] Gongalsky M B, Pereira A, Manankov A A, Fedorenko A A, Vasiliev A N., Soloviev V V, Kaudryavtsev A A, Sentis M, Kabashin A V, Timoshenko V Yu 2016 Laser-synthesized oxide-passivated bright Si quantum dots for bioimaging, *Sci. Rep.* **6** 24732.
[4] Osminkina L A, Timoshenko V Yu 2016 Porous silicon as a sensitizer for biomedical applications: Mini-review. Mesopor. *Biomat.* **3** 39–48.
[5] Tolstik E, Osminkina L A, Matthäus C, Burkhardt M, Tsurikov K E, Natashina U A, Timoshenko V Yu, Heintzmann R, Popp J, Sivakov V 2016 Studies of silicon nanoparticles uptake and biodegradation in cancer cells by Raman spectroscopy, *Nanomedicine: Nanotechnol., Biol. & Medic.* **12** № 7 1931-1940.
[6] Osminkina L A, Nikolaev A L, Sviridov A P, Andronova N V, Tamarov K P, Gongalsky M B, Kudryavtsev A A, Treshalina H M, Timoshenko V Yu 2015 Porous silicon nanoparticles as efficient sensitizers for sonodynamic therapy of cancer, *Micropor. Mesopor. Mat.* **210** 169-175.
[7] Sviridov A P, Osminkina L A, Kharin A. Yu, Gongansky M B, Kargina J V, Kudryavtsev A A, Bezsudnova Yu I, Perova T S, Geloen A, Lysenko V, and Timoshenko V Yu 2017 Cytotoxicity control of silicon nanoparticles by biopolymer coating and ultrasound irradiation for cancer theranostic applications, *Nanotechnol.* **28** (10) 105102.
[8] Timoshenko V Yu, Kudryavtsev A A, Osminkina L A, Vorontsov A S, Ryabchikov Yu V, Belogorokhov I A, Kovalev D, and Kashkarov P K 2006 Silicon nanocrystals as photosensitizers of active oxygen for biomedical applications, *JETP Lett.* **83**(9) 423-426.
[9] Osminkina L A, Tamarov K P, Sviridov A P, Galkin R A, Gongalsky M B, Solovyev V V, Kudryavtsev A A, Timoshenko V Yu 2012 Photoluminescent biocompatible silicon nanoparticles for cancer theranostic application, *J. Biophoton.* **5** 529–535.
[10] Hong Ch, Lee J, Zheng H, Hong S-S and Lee Ch 2011 Porous silicon nanoparticles for cancer photothermotherapy. *Nanosc. Res. Lett.* **6** 321.
[11] Tamarov K P, Osminkina L A, Zinovyev S V, Maximova K A, Kargina J V, Gongalsky M B, Ryabchikov Yu, Al-Kattan A , Sviridov A P, Sentis M, Ivanov A V, Nikiforov V N, Kabashin A V, and Timoshenko V Yu 2014 Radio frequency radiation-induced hyperthermia using Si nanoparticle-based sensitizers for mild cancer therapy, *Sci. Rep.* **4** 7034.
[12] Gongalsky M B, Kargina Yu V, Osminkina L A, Perepukhov A M, Gulyaev M V, Vasiliev A N, Pirogov Yu A, Maximychev A V and Timoshenko V Yu 2015 Porous silicon nanoparticles as biocompatible contrast agents for magnetic resonance imaging, *Appl. Phys. Lett.* **107**: 233702.
[13] Schmidt V, Wittemann J V, Senz S and Gösele U 2009 Silicon nanowires: a review on aspects

IOP Conf. Series: Journal of Physics: Conf. Series **945** (2017) 012002 doi:10.1088/1742-6596/945/1/012002

of their growth and their electrical properties, *Adv. Mater.* **21** 2681–2702.

[14] Rodichkina S P, Osminkina L A, Isaiev M, Pavlikov A V, Zoteev A V, Georgobiani V A, Gonchar K A, Vasiliev A N and Timoshenko V Yu 2015 Raman diagnostics of photoinduced heating of silicon nanowires prepared by metal-assisted chemical etching, *Appl. Phys. B: Lasers & Optics* **121** (3) 337-344.

[15] Gaisler S V , Semenova O I, Sharafutdinov R G and Kolesov B A 2004 Analysis of Raman spectra of amorphous-nanocrystalline silicon films. *Phys. Solid State,* **46**, pp. 1528–1532.

[16] Richter H, Wang Z P and Ley L 1981 The one phonon Raman spectrum in microcrystalline silicon. *Solid State Comm.* **39** 625–629.

[17] Maslova N E, Antonovsky A A, Zhigunov D M, Timoshenko V Yu, Glebov V N, Seminogov V N 2010 Raman studies of silicon eanocrystals embedded in silicon suboxide layers. *Semiconductors* **44** 1040–1043.

Morphological and immunological criteria of minimal residual disease detection in children with B-cell precursors acute lymphoblastic leukemia

O A Beznos[1], L Yu Grivtsova[1], A V Popa[1], M A Shervashidze[1], I N Serebtyakova[1], N N Tupitsyn[1], V U Selchuk[2, 3], O P Grebennikova[3], G V Titova[3]

[1]N.N. Blokhin Russian Cancer Research Center, Kashirskoe shosse 23, 115478, Moscow, Russia
[2]National Research Nuclear University MEPhI (Moscow Engineering Physics Institute), Kashirskoe shosse 31, 115409, Moscow, Russia
[3]A.I. Evdokimov Moscow State University of Medicine and Dentistry, Delegatskaya ulica, 20/1, 127473, Moscow, Russia

e-mail: beznos.olga@gmail.com

Abstract. One of the key factors of prognosis and risk stratification in patients with B-cell precursor acute lymphoblastic leukemia (BCP-ALL) is minimal residual disease (MRD). Identification of MRD on the day 15th is one of the most significant in prognosis of the disease. We compared data of a morphological and flow cytometry results of assessment of a bone marrow (BM) at the day 15th of induction chemotherapy in children with BCP-ALL.

1. Introduction

Detection of minimal residual disease (MRD) in bone marrow (BM) in children with B-cell precursor acute lymphoblastic leukemia (BCP-ALL) is one of the key factor of prognosis and at each step of the therapy it is solved particular clinical task [1].

Long time the assessment of therapy efficacy was based on a morphological method – light microscopy of a bone marrow sample stained by Romanovsky-Gimze method with counting percentage of blast cells.

According to morphological criteria on the level of blasts allocates M-variants: M1 - <5% of blast cells, 5-25% - M2 and more then 25% of blast cell – M3 [3]. Despite the relative simplicity of execution and economic appeal of this method, it's sensitivity is low (up to 102). Also certain difficulties is made by a mimicry of blasts and impossibility to distinguish them from the regenerating BM precursors on later terms of MRD monitoring that reveals limitation of morphological approach, and needed of search of alternative techniques.

Multicenter international research works proved clinical significance of immunological MRD detection by flow cytometry (FC) [4]. Morden FC has comparable sensitivity with molecular methods due to the opportunity of assessment up to 12 parameters at each cell and analyzing millions of cells at the same time [9]. Now monitoring of MRD baised on detailed charachteristics of blast cells immunophenotype and selection of optimal combinations of antigens at the stage of diagnosis [10].

One of the most significant time point in MRD detection is 15th day of induction chemotherapy, with reflects a primary answer of tumor to treatment [4-6]. Taking into account the lack of normal BCP in BM at the day 15 [4, 5] FC strategy at this time point is to detect $CD10^+$BCP in case of pre-pre-B immunosubtype. Earlier, we proved reliable comparability of 3- and 4-5 colors FC in MRD detection at the day 15.

2. Materials and methods

In a research were enrolled 130 children (64 boys and 66 girls) with BCP-ALL, whom MRD were detected at the day 15 of induction chemotherapy. Middle age was 5.5 (1-17 years). Pre-pre-B immunosubtype was diagnosed in 92.3%, pro-B – 7.7%. Patients were treated according to ALL IC BFM 2002 (40 of 130) and ALL IC BFM 2009 (90 of 130) protocols. In all cases morphological and immunological tests were provided from one tube with BM aspirate. In 120 cases MRD detection was with 3-5 colors FC protocol. In 8 cases were use 8-color EuroFlow FC protocol.

Immunological assessment of MRD was based on detection of CD10 and CD34 expression and CD45 dim expression or its lack in a gate of $CD19^+$ cells. In all cases to exclude debris and to simplify calculation was used Syto-family dye (Syto16 or Syto41).

The algorithm of a gating is presented at figure 1.

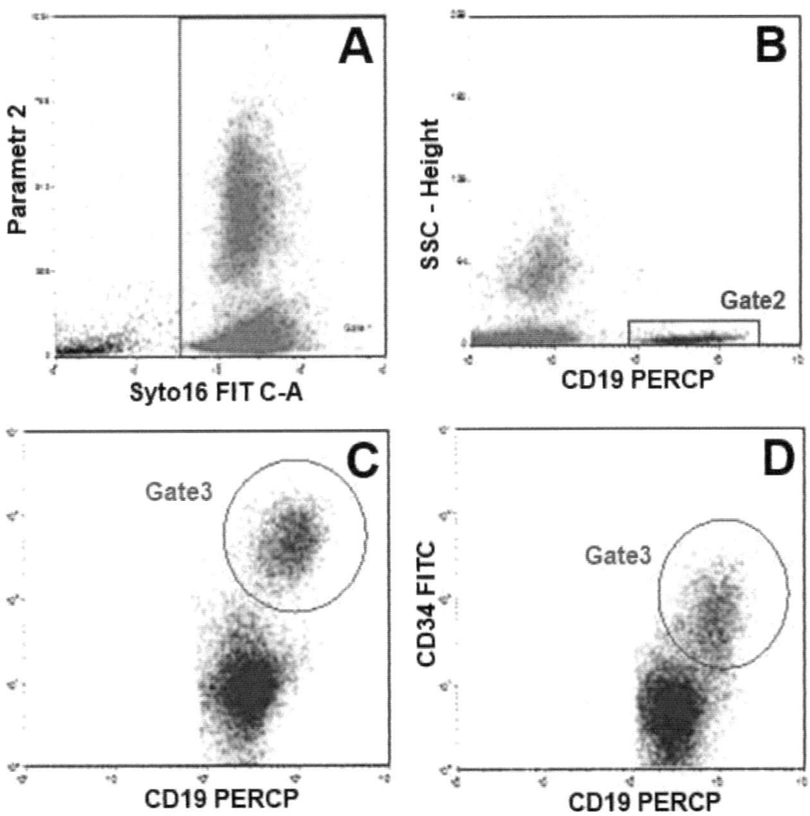

Figure 1. The algorithm of gating in MRD detection at the day 15 of induction chemotherapy. At the cytogramm A performed calculation of nucleated cells according to the expression Syto16 (x axis) vs side scatter parameters (y axis) – gate 1, red. $Syto16^+$ 91.65%. Cytogramm B performed detection of $CD19^+$ cells 11.81% (CD19 – x axis vs side scatter parameters (y axis) – gate 2, blue. Cytogramm C – 2.52% $CD10^+$BCP detected by expression of CD10 antigen (y axis) vs CD19 (x axis), gate 3, green. Cytogramm D – 2.35% $CD34^+$ BCP detected by expression CD34 antigen (y axis) vs CD19 (x axis), gate 3, green.

In case of pro-B ALL immunosubtype MRD detection based on calculation of $TdT^{+}cyCD22^{+}$BCP (figure 2).

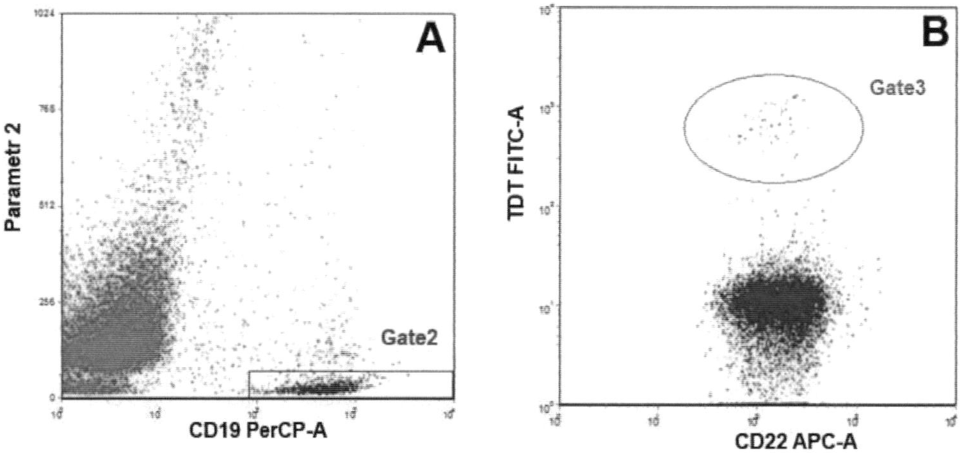

Figure 2. MRD detection by expression of TdT and cyCD22 antigens in case of pro-B ALL. Cytogramm A – calculation of $CD19^{+}$ cells (CD19 – x axis vs SSC parameters – y axis) gate 2 – 5.5%, blue. Cytogramm B – 0.03% of BCP detected (TdT y axis vs cyCD22 x axis), gate 3, green.

3. Results

Samples were analyzed according to criteria of M-variants. M1 group was made by 80% of samples, M2 group – 14.6%. In M3 group was 5.4% of samples. In each group the quantity of blast and lymphocytes, cellularity of samples, and also the MRD levels were compared. Morphological criteria are compared with immunological data. Each group of samples is in details analyzed concerning the MRD levels according to criteria of stratification of the BFM protocol: 0.1% - standard risk, 0.1-10.0% - intermediate risk and 10,0% and more – high risk group. Standard threshold level of MRD-negativity is 0.01% of tumor cells in a sample.

M1 group samples (104 samples) was characterized by the higher percent of lymphocytes (28.0%). Presence of MRD-positive samples (FC) at the M1 group (group of the good morphological answer) can be explained with a mimicry the blast cells, that is their full morphological similarity with lymphocytes. It complicates a possibility of accurate morphological verification of residual leukemic blasts.

M2 group was made by 19 BM aspirates. BM samples with the maintenance of MRD cells less than 0.1% (standard risk) according to FC in this group were absent. Morphological and FC date and the M2 group, didn't contradict each other.

In M3 group was made by 7 samples. In the group there was no samples with quantity of MRD cells less than 1.0% according to a FC. In 85.7% of samples of this group (6 of 7) on data of FC has revealed 10.0% and more MRD-positive cells.

By comparison of average values of indicators of the analyzed groups reliable distinction in number of lymphocytes is established when comparing the M1 and M3 groups ($p < 0.0001$) and the M2 and M3 groups ($p = < 0.0001$).

4. Conclusion

Thus, from the carried-out work it is possible to draw the following conclusions.

At the MRD monitoring it is necessary to provide morphological and immunological tests using BM aspirate of one tube with obligatory calculation of sample cellularity.

Morphological and FC data are coincide. In the M2 and M3 groups, immunological MRD-negative samples aren't revealed. In the M1 group the majority of samples were MRD-positive according to FC, 42.9% from them contained no more than 0.1% of MRD cells.

Presence the MRD-positive samples at M1 group can be explained with a mimicry the blast cells, that is their similarity with lymphocytes that complicates accurate morphological verification of blasts.

FC, having higher sensitivity in comparison with a morphological method, allows to stratify more precisely patients on risk groups (at the M1 group there were samples both from group standard and intermediate and from group of high risk).

Introduction to the panel of MRD detection the dyes of the Syto-family is obligatory. It allows to allocate accurately nucleated cells of sample, simplifies recalculation of quantity of MRD cells and gives the chance to exclude debris from the analysis. Dilution of a sample by debris can lead to understating of quantity of MRD cells and as a result – incorrect risk stratification of patients.

Acknowledgments
This work was supported by Competitiveness Program of National Research Nuclear University MEPhI.

References
[1] Borowitz M J *et al.* 2008 *Blood* **111**(12) 5477-85
[2] Dworzak M N *et al.* 2002 *Blood* **99**(6) 1952-58
[3] Basso G *et al.* 2009 *Jornal of clinical oncology* **27**(31) 5168-74
[4] Coustan-Smith E *et al.* 2002 *Blood* **100**(1) 52-58
[5] Coustan-Smith E *et al.* 2006 *Blood* **108**(1) 97-102
[6] Schrappe M *et al.* 2000 *Blood* **95**(11) 3310-3322
[7] Grivtsova L U *et al.* 2008 (in Russia) *Haematopoiesis immunology* 2 8-33
[8] Grivtsova L U *et al.* 2011 (in Russia) *Haematopoiesis immunology* 1 35-54
[9] van Dongen J J *et al.* 2015 *Blood* **125**(26) 3996-4009
[10] Beznos O A *et al.* 2017 (in Russia) *Oncohematology* **10**(2) 158-168

G-CSF-primed autologous and allogeneic bone marrow for transplantation in clinical oncology. Cell content and immunological characteristics

L Yu Grivtsova[1], K N Melkova[1], N A Kupryshkina[1], I K Vorotnikov[1], T A Grigoryeva[1], V Yu Selchuk[2,3], O P Grebennikova[3], G V Titova[3], N N Tupitsyn[1]

[1]N.N. Blokhin Russian Cancer Research Center, Kashirskoe shosse 23, 115478, Moscow, Russia
[2]National Research Nuclear University MEPhI (Moscow Engineering Physics Institute), Kashirskoe shosse 31, 115409, Moscow, Russia
[3]A.I. Evdokimov Moscow State University of Medicine and Dentistry, Delegatskaya ulica, 20/1, 127473, Moscow, Russia

e-mail: grivtsova@mail.ru

Abstract. 60 samples of G-CSF-primed bone marrow (39 cancer patients and 21 healthy donors) to be used for transplantation to cancer patients were analyzed and compared by main characteristics with historical control and 13 bone marrow samples from control patient with mastopathy. Basing on morphological and multicolor flow cytometry findings certain characteristics of G-CSF-primed bone marrow were discovered, such as a significant increase in blast count in cancer patients as compared to donors and control patients ($p<0.037$), a higher neutrophil maturation index ($p<0.001$) and a lower percentage of mature lymphocytes ($p<0.008$) as compared to the control group. Among lymphocyte populations G-CSF-priming was associated with a significant increase in the total of mature CD3+ T-cells and CD8+ T-killers ($p<0.0001$) and a decrease in CD56+CD3- and/or CD16+CD3- NK-cells ($p<0.006$) both in cancer patients and healthy donors in comparison with the controls.

1. Introduction

Allogeneic hemopoietic tissue transplantation alone is a therapy of choice in many patients with hematology malignancies [1, 2, 3, 4].

Bone marrow is preferred for allogeneic transplantation mainly due to lower (4 to 10-fold) content of mature T-cells involved in graft versus host reaction (GVHR) [5 – 9]. Donors' T-cells though mediate GVHR are needed for adequate recovery of recipient's immunity functioning and, what is of principal importance, some of their subsets (CD8+ memory T-cells) are mediators of graft versus tumor reaction [10 – 13].

There is a minor regulatory T-cell subset predominating in bone marrow that acts as a protector of post-transplantation GVHR [14, 15], in spite of the fact that most of these cells are mediators of GVHR in allogeneic transplantation of mobilized PB [16].

So, from the immunological standpoint bone marrow is a more attractive material for both auto- and allotransplantations in cancer patients. It is of much importance to study in detail characteristics of stimulated bone marrow due to high clinical need in this material and equivocal role of some lymphocyte subsets.

The purpose of this study was to analyze in detail morphological and immunological (lymphocyte subset composition) characteristics of granulocyte colony stimulating factor (G-CSF)-primed bone marrow from cancer patients and healthy donors and to compare characteristics of the stimulated bone marrow samples with normal parameters.

2. Materials and methods

We analyzed 39 stimulated bone marrow samples from 39 adult patients with hematology malignancies taken for autologous transplantation and 21 granulocyte colony stimulating factor (G-CSF)-primed bone marrow samples taken from 21 healthy donors to transplant to adult patients with hematology malignancies.

The cancer patient group included 22 females and 17 males, mean age was 38 (range 18 to 57) years. Most patients had Hodgkin's lymphoma (28) and 11 patients had acute myeloid leukemia (AML). The bone marrow donor group consisted of 10 men and 11 women, mean age was 40 and median age was 37 (range 18 to 61) years.

Both autologous and allogeneic bone marrow samples were taken after G-CSF-priming (at 100 mcg/kg daily) as performed by similar schedules in the patients and donors. The control group was composed of 13 bone marrow aspiration biopsies from females with breast cancer diagnosis excluded.

Each bone marrow sample was characterized by morphology and immunology, analyses were made from the same tube, smears and cells for immunological analysis were prepared directly at the laboratory. Cellularity of samples was counted using a Micros 60 hematology analyzer.

Lymphocyte subset composition and the number of precursor cells were studied immunologically (multicolor, 4- to 8-color flow cytometry) using direct conjugates of monoclonal antibodies to CD34 precursor-cell antigen (Figure. 1), and to T-, B- and NK-cell main antigens (CD3, CD4, CD8, CD16, CD56, CD94). Hemopoietic stem cell count was made within all nucleated cells (modified protocol for stem cell count by flow cytometry) [17]. Lymphocyte subsets were counted within the lymphocyte gate (figure 2).

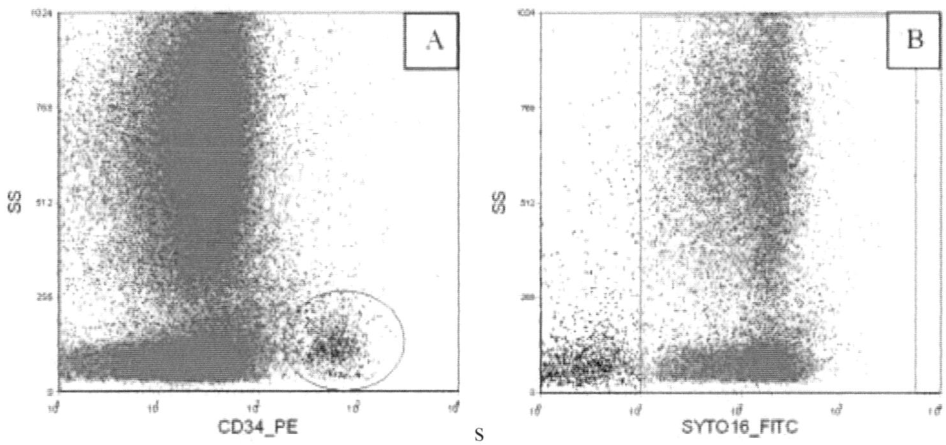

s

Figure 1. Assessment of CD34+ stem cell count in a donor's bone marrow sample. A – assessment of proportion of nucleated cells using a *syto 16* nucleotropic dye. B – assessment of stem cell (CD34+) count within nucleated (syto16+) cells in the sample.

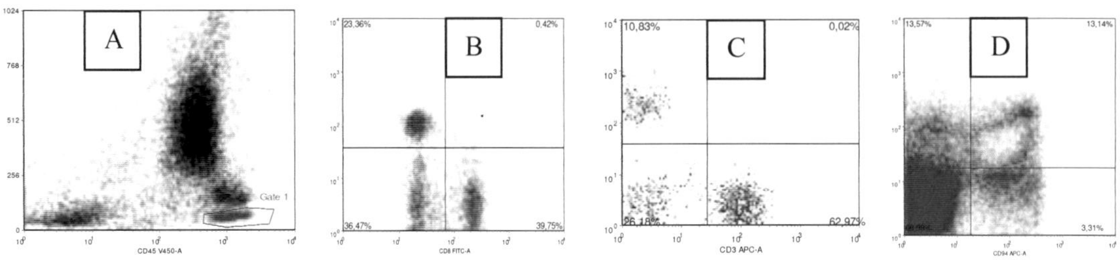

Figure 2. Cytogram A shows the lymphocyte gate CD45+ (blue). Cytograms Б, В,Г demonstrate lymphocyte subset composition: Figure B – B-cell proportion is 10.83% of lymphocytes (*y*-axis, upper left quadrant), T-lymphocytes (CD3+) are 62.9% (*x*-axis), Figure C — CD4+ T-helper proportion is 23.4% (*y*-axis, upper left quadrant), CD8+ T-killer portion is 39.7% (*x*-axis, lower right quadrant), Figure D — most (13.2% of lymphocytes) of NK-cells (yellow) are a CD56+CD94+ subset

Statistical analysis of data was made using an SPSS 17.0 software.

3. Results

Comparison of bone marrow morphological composition demonstrated a significantly higher blastosis ($p=0.012$ and $p=0.037$) and lower lymphocyte proportion ($p=0.017$ and $p<0.0001$) in bone marrow samples from cancer patients than in either donors' or control bone marrow samples. Neutrophil maturation index was significantly higher in cancer patients as compared to donors' stimulated bone marrow and control samples ($p=0.026$ and $p<0.0001$, respectively).

Mature T-cells (CD3+) were predominant among lymphocytes in stimulated bone marrow with most of them being CD8+ T-killers (table 3).

Comparison of immunological characteristics of bone marrow cells demonstrated predominance of CD34+ precursors in cancer patients' samples as compared to the donors' ones though the difference was not significant. The largest number of CD34+ cells (10.9%) was found in AML cases, and the number of CD34+ cells was proportional to blast count ($p<0.0001$) in cancer patients, while no such correlation was seen in the donors' group.

Stimulated bone marrow samples from cancer patients showed a significant predominance in percentage of CD3 lymphocytes ($p=0.04$), CD8+ T-killers ($p<0.0001$) and accordingly lower immunity regulating index ($p<0.0001$) as compared to donors' samples.

Stimulated bone marrow samples from cancer patients demonstrated a reverse relationship between immunity regulating index and neutrophil maturation index ($R= - 0.363$, $p=0.027$), this correlation in the donors' group was not significant.

Comparison of lymphocyte subset composition in bone marrow samples from HL versus AML subgroups showed a significant predominance of CD3+ lymphocytes (82.9±3.3% vs. 68.3±4.5%; $p=0.021$) and CD4 leukocytes (2.14±0.2% vs. 1.02±0.16%; $p=0.001$) in the AML samples.

Samples of stimulated bone marrow from HL patients were characterized by a significantly higher content of CD56+ NK-cells (5.6±0.5% vs. 3.5±0.5%, $p=0.05$) and a higher proportion of CD19+ B-lymphocytes (26.4±5.9% vs. 8.1±2.9%, $p=0.01$) as compared to AML subgroup.

Comparison of immunological characteristics of stimulated bone marrow samples (from both cancer patients and donors) with those from the controls demonstrated a significantly higher proportion of T-killers and the lowest immunity regulation index in samples from cancer patients as compared with the controls ($p=0.0001$). Stimulated bone marrow lymphocytes from both cancer patients and donors contained a significantly higher proportion of CD16+ and CD56+ NK-cell subsets ($p<0.06$ and $p<0.004$, respectively) as compared with the control group.

4. Conclusion

Our study discovered statistically significant differences in morphological and immunological (lymphocyte subsets) compositions of G-CSF-primed bone marrow as increase in blasts, predominance of younger granulocytes due to promyelocytes and myelocytes, increased proportion of mature erythroblasts among erythroid cells and decreased proportion of mature lymphocytes in samples from both cancer patients and donors. As concerns lymphocyte subsets, G-CSF-priming was associated with predominance of the total of mature CD3+ T-cells and CD8+ T-killers and decreased number of CD56+CD3- and/or CD16+CD3- NK-cell subsets.

Acknowledgments

This work was supported by Competitiveness Program of National Research Nuclear University MEPhI.

References

[1] Appelbaum F R 2007 *The New England Journal of Medicine* **357**(15) 1472–1475
[2] Baldomero H *et al.* 2011 *Bone Marrow Transplantation* **46**(4) 485–501
[3] Pasquini M C and Zhu X 2014 *Center for International Blood & Marrow Transplant Research Conference* Available from: http://www.cibmtr.org
[4] Morton J, Hutchins C and Durrant S 2001 *Blood* **98**(12) 3186-3191
[5] Flowers M E *et al.* 2002 *Blood* **100**(2) 415-419
[6] Storek J *et al.* 1997 *Blood* **90**(12) 4705-4709
[7] Holtick U *et al.* 2014 *Cochrane Database Syst Rev* **4** CD010189
[8] Anasetti C *et al.* 2012 *New England Journal of Medicine* **367**(16) 1487-1496
[9] Eapen M *et al.* 2015 *Journal of Clinical Oncology* **33**(4) 364-369
[10] Mackall C L and Gress R E *et al.* 1997 *Immunological reviews* **157**(1) 61–72
[11] Collins R H *et al.* 1997 *Journal of clinical oncology* **15**(2) 433–444
[12] Kolb H J *et al.* 1995 *Blood* **86**(5) 2041–2050
[13] Zheng H *et al.* 2009 *The Journal of Immunology* **182**(10) 5938 –5948
[14] Zhao X Y *et al.* 2015 *Journal of Translational Medicine* **13**(1) 145
[15] Rezvani K *et al.* 2006 *Blood* **108**(4) 1291–1297
[16] Pabst C *et al.* 2007 *Clinical Cancer Research* **13**(10) 2916–22
[17] Grivtsova L Yu and N N Tupitsyn (in Russia) *Medical alphabet* **3**(30(293)) 50-53

Optimization of ^{31}P magnetic resonance spectroscopy *in vivo*

A V Manzhurtsev[1,2,4], T A Akhadov[1] and N A Semenova[1,2,3]

1. Radiology Department, Research Institute of Childen Emergency Surgery and Trauma, 22 Bolshaya Polyanka Street, Moscow 119180, Russia
2. Enzyme Catalysis Kinetics Department, N.M. Emanuel Institute of Biochemical Physics of RAS, 4 Kosygina Street, Moscow 119334, Russia
3. Chemical and Biologycal Processes Dynamics Department, N.N. Semenov Institute of Chemical Physics of RAS, 4 Kosygina Street, Moscow 119334, Russia
4. Medical Physics Department, National Research Nuclear University "MEPhI", 31 Kashirskoe Shosse, Moscow 115409, Russia

Andrey.man.93@gmail.com

Abstract. The main problem of magnetic resonance spectroscopy on phosphorus nuclei (^{31}P MRS) is low signal-to-noise ratio (SNR) of spectra acquired on clinical (3T) scanners. This makes quantitative processing of spectra difficult. The optimization of method on a single-voxel model reported in current work implicates an impact of the spin-lattice (T_1) relaxation on SNR, and also evaluates the effectiveness of Image Selected InVivo Spectroscopy (ISIS) pulse sequence modification for the increase of SNR.

1. Introduction

Magnetic resonance spectroscopy on ^{31}P nuclei (^{31}P MRS) is a unique way of non-invasive investigations of energy and lipid metabolism of human and animal tissues *in vivo*, as well as a method of intracellular pH measurement. ^{31}P MRS has demonstrated its high effectiveness in metabolic processes investigations in the norm and in pathology, including the studies of energy metabolism response to neurostimulation in local brain zones [1].

The ^{31}P MR spectra of brain display signals for the phosphate groups of energy metabolism participants: phosphocreatine (PCr), ATP (3 peaks: b-, a- and g-ATP), inorganic phosphate (Pi); dinucleotides (DN) and the signals of lipid metabolism participants: phosphomonoesters (PME) phosphocholine (PC) and phosphoethanolamine (PC), phosphodiesters (PDE) glycerophosphocholine (GPC), glycerophosphoethanolamine (GPE) and unidentified (GPX). From the Pi frequency measured relative to PCr, one can calculate the intracellular pH value [2].

When comparing with protons, phosphorus nuclei have ~2.5 lower gyromagnetic ratio. This creates the main ^{31}P MRS in vivo problem: the necessity of increasing the volume of interest and spectra aquisition time to achieve satisfactory SNR values for the metabolites of interest. Such parameters of spectra obtaining like time of repeat (TR) and number of signals averaging (NSA) provide SNR gain by means of aquisition time increase. The TR parameter optimization for each particular study, as well as proton-phosphorus decoupling and nuclear Overhauser enhancement (NOE) allow to decrease the required time.

The aim of current study is to obtain ^{31}P MR spectra with high SNR at minimal time of acquisition.

2. Materials and Methods

Eleven healthy subjects (6 male, 5 female) aged 18-50 became participants of this study. All subjects passed a standard diagnostic MRI examination at Philips Achieva 3.0T, including T_2-, T_1-, diffusion-weighted and FLAIR images. No structural abnormalities according to MRI data served as the study entry criterion.

The spectroscopy study was performed by using a transmit-receive head dual-tuned 31P/1H bird cage coil by Rapid Biomed, Germany. Keeping in mind the chemical shift displacement artifact [3], the spectroscopy volume of interest (voxel) sized 80x60x60 mm was located in such a way as to contain the most part of brain tissue. Voxel location is shown in figure 1. Voxel positioning was performed with the help of 3D localizer (survey) images and a set of 2D axial T_2-weighted images.

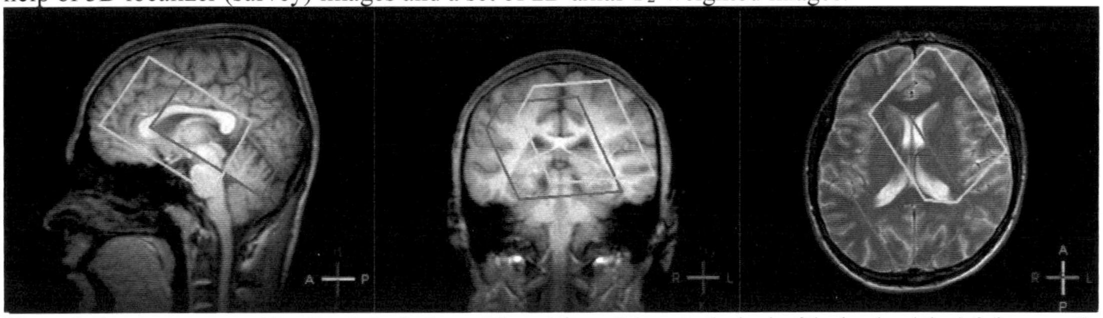

Figure 1. Spectroscopy voxel location in brain tissue. The red frame (closer to the back of the head and the left frame on the middle image) displays the volume from which creatine phosphate (PCr) signal is registered, the cyan frame (closer to the lobe) at the same time displays the source of beta phosphate of ATP molecule signal (b-ATP).

Pulse sequence (PS) ISIS was used for voxel localization, the parameters were as follows: echo time (TE)=0.1 ms, total magnetization vector flip angle (FA)=35°, number of spectral points=1024, spectral bandwidth (BW)=4000 Hz, number of *dummy* scans for steady-state magnetization settlement=2. For SNR increasing PS ISIS was modified with broadband proton irradiation for decoupling implementing during ^{31}P signal registration via PS waltz 4 [4], and broadband proton irradiation before 31P excitation pulses for NOE implementing via PS waltz 16 [4].

The ^{31}P MR spectra SNR optimization study consisted of two parts. In the first part a standard PS ISIS without proton irradiation at any time was used for 3 spectra obtaining with the same NSA=64 and TR values 2000, 3000 and 4000 ms. In the second part TR=3000 ms and NSA=64 were fixed, decoupling and NOE were used. Eleven spectra were obtained with varying NOE parameter *mixing (mix) time*, equal to 0, 250, 500, ..., 2500 ms. The total length of the study was ~45 minutes, which is acceptable for a person in MRI scanner.

Spectral processing was performed in Spectroview program. In both parts of the study, SNR of each resonance in spectrum was measured. In the first part, the SNR increase (in %) for each spectral line of each subject was calculated when TR was increased from 2000 ms to 3000 ms, from 3000 ms to 4000 ms and from 2000 ms to 4000 ms. In the second part, the SNR of each peak for different *mix times* were normalized on the corresponding values at *mix time*=0

Statistical analysis was performed in STATISTICA 12 using Mann-Whitney U-criterion to evaluate statistical significance of differences between groups.

3. Results

The SNR changes (in %) for all ^{31}P spectrum peaks for each TR pair of the first part of the study are demonstrated in table 1. It is obvious that an increase of TR leads to the gain in SNR, but not for all metabolites and not for all pairs of TR this gain exceeds the one that could be achieved by increasing of NSA and using smaller TR.

The example of SNR dependencies (for b-ATP and Pi peaks) as a function of mix time is demonstrated on graphs, figure 2. The SNR increase with the growth of mix time is common for all metabolites. The SNR gain for ATP, PC, GPC, GPX, DN resonance lines slows down at mix time values

~1250-1750 ms. For PCr, Pi, GPE, PE graphs the SNR increase is observed up to 2250-2500 ms values of mix time. The maximal SNR gain by means of NOE is around ~30%.

Table 1. Average percentage increase in SNR (SD). * marks the statistically significant ($p<0.05$ according to Mann-Whitney criterion) increase, that exceeds the effectiveness threshold (see discussion).

Peak	TR=3000 ms vs TR=2000 ms Threshold = 18.1 %	TR=4000 мс vs TR=3000 мс Threshold = 13.4%	TR=4000 мс vs TR = 2000 мс Threshold = 41.4%
PCr	33.0 (4.7)*	11.2 (3.7)	44.8 (9.4)*
γATP	18.5 (5.7)*	7.1 (5.8)	27.5 (13.1)
αATP	16.4 (7.4)	6.7 (2.1)	29.6 (15.7)
βATP	15.9 (6.0)	7.2 (4.1)	32.1 (17.1)
Pi	35.3 (11.2)*	18.5 (6.6)*	44.1 (15.5)*
PE	30.1 (13.6)*	11.8 (10.2)	56.6 (19.2)*
PC	23.4 (7.1)*	4.1 (3.1)	42.1 (27.4)*
GPE	37.7 (17.9)*	20.2 (9.5)*	73.6 (28.4)*
GPC	29.3 (16.1)*	11.0 (8.8)	66.5 (31.0)*
GPX	15.0 (11.5)	11.3 (5.1)	37.9 (19.8)
DN	20.1 (5.8)*	9.4 (4.5)	22.3 (10.5)

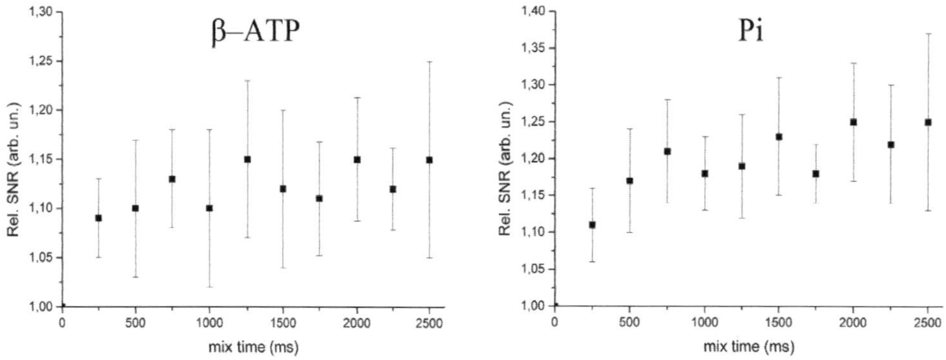

Figure 2. Relative SNR change as a function of mix time for all metabolites.

4. Discussion

The larger TR value, the smaller NSA must be to keep spectrum acquisition time constant. SNR is proportional to the square root of NSA, so, when coming to TR=3000 ms from TR=2000 ms NSA has to be decreased by 33,3%, that automatically decreases SNR by 18.1%. This value is the effectiveness threshold: if, when NSA is kept constant, SNR gain by means of TR increase from 2000 ms to 3000 ms exceeds 18.1%, the usage of larger TR is profitable. The effectiveness thresholds are calculated analogously in other cases: it is 13.4% when changing TR=3000 ms to 4000 ms, and 41.4% when TR=2000 ms changes to 4000 ms.

Table 1 demonstrates that TR = 3000 ms is more preferable than TR = 2000 ms: the SNR gain exceeds the threshold for most of metabolites. It is connected with the fact that the time required for the full relaxation of ^{31}P spectrum metabolites ($5T_1$) is significantly higher than TR=2000 ms [5]. A comparison of TR=4000 ms with TR=3000 ms shows that SNR gain for most of metabolites is not enough to compensate the loss in signal owing to NSA decrease. This statement is false for Pi and GPE peaks, and that agrees with the fact that T_1 values of Pi and PDE are the ones of most large.

TR=4000 ms comparison with TR=2000 ms demonstrates the effectiveness of the larger TR for groups of PME and PDE peaks, PCr and Pi, but not to ATP and DN peaks. For that reason TR=4000 ms should be used in case of necessity to provide the maximal SNR_{Pi} and when lipid metabolism is of interest in the study. SNR_{Pi} is crucial for intracellular pH determination, because the variation range of this parameter *in vivo* is rather narrow [6]. Local brain pH changes owing to no damaging exposure, for example, neuronal activation, are ~1% [7], hence, may be missed because of low SNR_{Pi}.

In summary, TR=3000 ms is universally applicable. That contradicts [8], where authors state that SNR of 3T brain ^{31}P spectra is insufficient when TR<4000 ms, but, unfortunately, no quantitative analysis of results is shown. We show that change of TR from 3000 ms to 4000 ms does not give profitable SNR gain for most of metabolites, although it is recommended in some cases. What is more, for the alone energy metabolism analysis TR≤2000 ms with not much increase of NSA may be used, that allows to shorten the study time. For example, in the study of energy metabolism response on visual cortex stimulation in schizophrenia patients [1], where the fastest data acquisition was required, we revealed energy metabolism alterations using TR=1200 ms.

The larger TR, the longer it is possible to use proton irradiation before ^{31}P excitation pulse. By means of NOE during this excitation the difference increases between populations of ^{31}P nucleus energy levels. The maximal value of heteronuclear NOE theoretically achieved is half of gyromagnetic ratios ratio, which means ~150% [9]. But in reality it is impossible to achieve such values: NOE enhancement monotonically rises with the *mix time* increase until the plateau, meaning equilibrium condition [10].

The time of the plateau beginning is strongly dependent on system parameters, particularly on participating nuclei and constant magnetic field strength [10]. According to the graphs on figure 2, noticeable SNR gain for most of metabolites ends at ~1500 ms of *mix time*. PCr, Pi, PE, GPE are exceptions, the increase is going on up to ~2500 ms. Maximal *mix time* is defined by TR parameter. At the universally applicable TR=3000 ms ~500 ms are required for ^{31}P signal registration, so *mix time* is constrained by ~2500 ms. This value is enough for a maximal SNR gain by means of NOE practically for all metabolites. When using TR=2000 ms for energy metabolism investigations, available maximal *mix time* ~1500 ms is enough for maximizing NOE on ATP resonance lines.

Acknowledgements
This work was supported by Competitiveness Program of National Research Nuclear University MEPhI.

References

[1] Manzhurtsev A, Semenova N, Ublinskii M, Akhadov T and Varfolomeev S 2016 *Rus. Chem. Bull., International Edition* **65** 1630–1636
[2] Petroff O A, Prichard J W, Behar K L, Alger J R, den Hollander J A and Shulman R G 1985 *Neurology* **35** 781–8.
[3] Kolkovsky A L L 2015 1H and 31P NMR Spectroscopy for the study of brain metabolism at Ultra High Magnetic Field from Rodents to Men. Universit´e Paris Sud - Paris XI Chapter 1.2.4 p. 39
[4] Shaka A J, Keeler J, Frenkiel T and Freeman R 1983 *J. Magn. Reson* **52** 335–338
[5] Ren J, Sherry A D and Malloy C R 2015 *NMR Biomed* **28** 1455–1462
[6] Chesler M 2003 *Physiol. Rev.* **83** 1183–1221
[7] Magnotta V A, Heo H Y, Dlouhy B J, Dahdaleh N S, Follmer R L, Thedens D R, Welsh M J, and Wemmie J A 2012 *PNAS* **109** 8270–8273
[8] Novak J, Wilson M, MacPherson L, Arvanitis T N, Davies N P and Peet A C 2014 *Eur. J. Rad* **83** e106– e112
[9] Claridge T D W 2016 *High-Resolution NMR Techniques in Organic Chemistry (Third Edition)* Elsevier UK
[10] Bertini I, Luchinat C, Parigi G 2001 *Solution NMR of Paramagnetic Molecules, Vol. 2, 1st Edition* Elsevier UK

IOP Conf. Series: Journal of Physics: Conf. Series **945** (2017) 012006 doi:10.1088/1742-6596/945/1/012006

The method of segmentation of leukocytes in information-measuring systems on the basis of light microscopy

V G Nikitaev, A N Pronichev, E V Polyakov, Yu V Zaharenko

National Research Nuclear University MEPhI (Moscow Engineering Physics Institute), Kashirskoe shosse 31, 115409, Moscow, Russia

E-mail: VGNikitayev@mephi.ru

Abstract. The paper considers the problem of leukocytes segmentation in microscopic images of bone marrow smears for automated diagnosis of the blood system diseases. The method was proposed to solve the problem of segmentation of contacting leukocytes in images of bone marrow smears. The method is based on the analysis of structure of objects of a separation and distances filter in combination with the watershed method and distance transformation method.

1. Introduction

Hematology analyzers and flow cytometry technique are widely used for analizes of blood and bone marrow aspirates in the diagnosis of diseases of the blood system [1-3]. Flow cytometry technique has advantages: high performance, the ability to define immunophenotypical subtype of acute leukemia. But it has the following disadvantages: the complexity of the analysis of the morphology of the cells, the inability to re-sample analysis, an incorrect classification of cells because of the significant variability of the measured characteristics of the cells [3-4].

Currently the automated analysis of microscopic images of peripheral blood and bone marrow smears is actively developing [5]. The procedure of detection of abnormal cells in such systems includes next steps: image enhancement, segmentation of cells, the description of the leukocytes in the language of signs, classification [1-17].

The problem of segmentation of cells in microscopic images of bone marrow smears has particular difficulty [11]. This is due to the fact that in these preparations the leukocytes are often in contact with each other– "stick together", forming chains and conglomerates. In this regard, task of developing segmentation algorithms that allows to separate "sticky" white blood cells is actual.

The purpose of the present paper is to describe results of development and research of a method of segmenting leukocytes on images of blood and bone marrow in large clusters of cells that stick together among themselves.

2. The leukocytes segmentation method

The study of segmentation methods were carried out on 200 microscopic images of bone marrow smears. Leukocytes were present in different quantities (1 to 45) in the one frame field.

Images were obtained by the system of computer microscopy(automated Olympus BX43 microscope with camera Imperx IPX-4M1ST-GCFB). Images were saved in BMP format, color-coded RGB24 (~16 million colors). The size of the image 1917x1092 pixels. The color value range of the nuclei by components: R=[47:122], G =[39:114], B=[95:143].

IOP Conf. Series: Journal of Physics: Conf. Series **945** (2017) 012006 doi:10.1088/1742-6596/945/1/012006

Examples of leukocytes in contact with each other are shown on Figure 1.

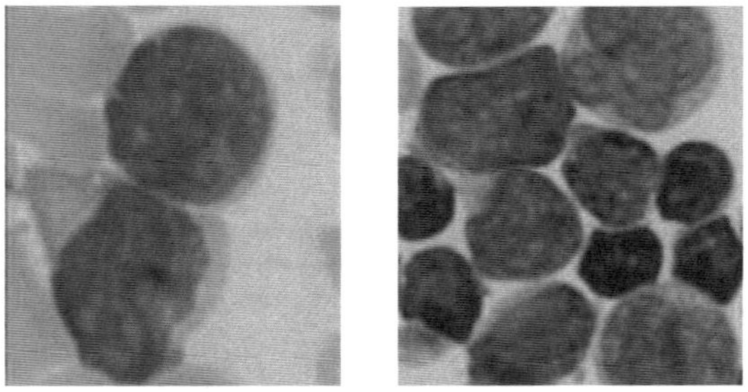

Figure 1. An example images of leukocytes in contact with each other. Images were obtained from bone marrow smears.

Cells separation algorithms were analyzed in according to the publications [6-8] in the first stage of solving of segmentation problem. For example, separation occurs by partitioning of the circuit on the connecting line between the angle points in the algorithms of angles finding. The downside of this algorithm is the case of angles greater than two (Figure 2)

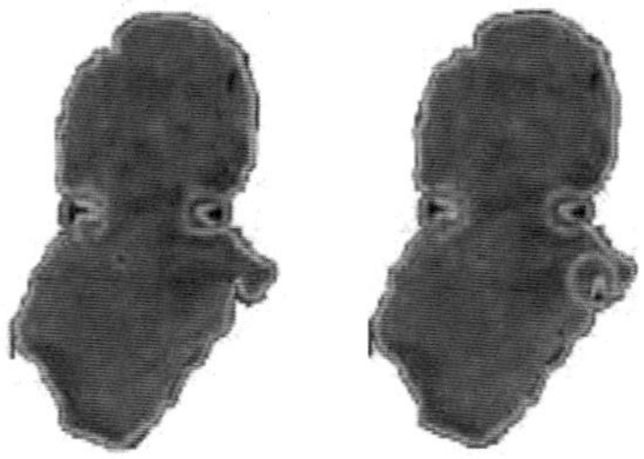

Figure 2. An example of the separation of leukocytes by finding the angles and the problems associated with separation at the corners.

One of the solution of the adherent leukocytes problem is application of algorithms of transformation of distances and algorithm of watersheds. This method is suitable for separating multiple leukocytes, since the watershed algorithm can detect all areas in the transformed image (Figure 3) [7-8]. The disadvantage of this approach is that there are the presence of artifacts, segmentation of voids inside the cells and space in the case of large concentrations of leukocytes. Finding local maxima was used for this problem decision, but that solution spoils the boundaries of the object. In this regard we have proposed an approach for the modernization of the method described above with the addition of filter cavities based on the method of labelling and filtering of the distances in the algorithm of transformation distances.

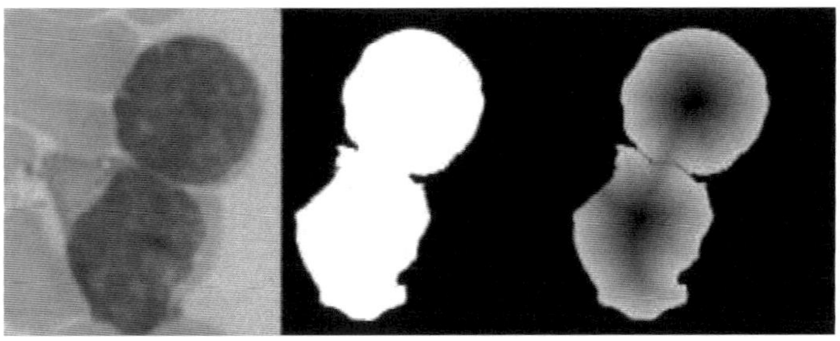

Figure 3. An example of results of methods of the watershed and transformation of distances for the separation of leukocytes.

A software module with a graphical user interface for computer image processing was developed in C++ with use of Qt library. That allows you to upload images and apply automatic analysis, and display results. An example of applying separation of leukocytes is shown in Figure 4 [9].

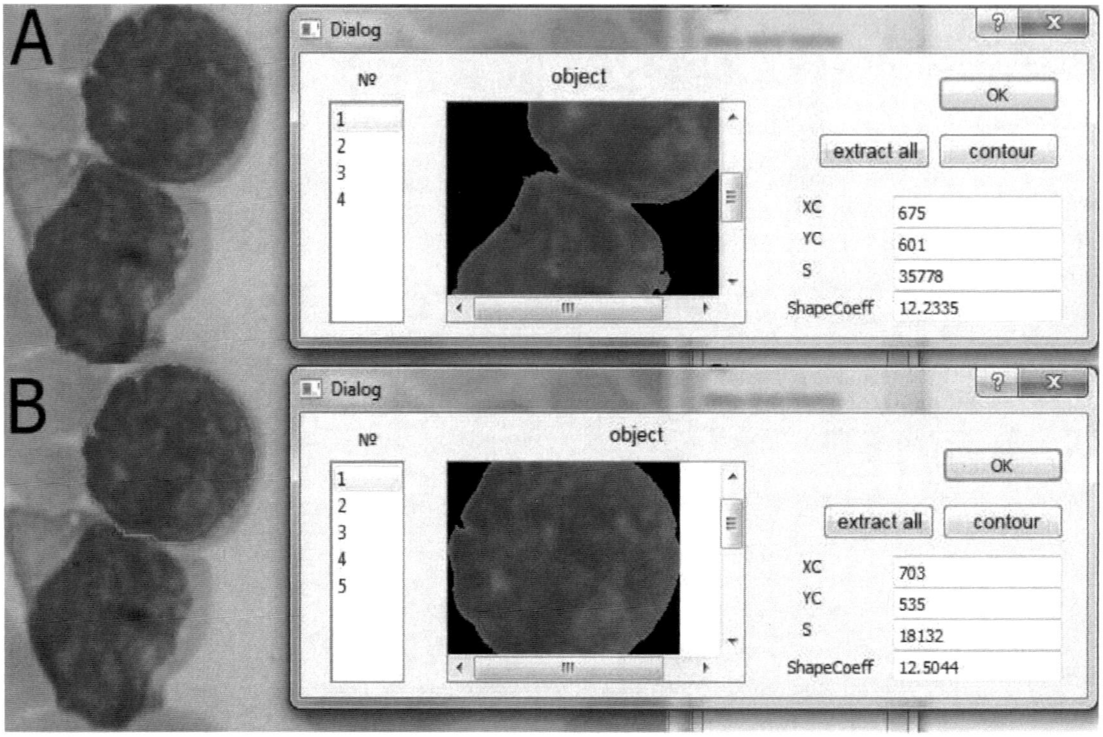

Figure 4. An example of segmentation of leukocytes in conditions of contact between the cells without the proposed method of segmentation (A) and using the proposed method (B).

The number of leukocytes was counted manually and programmatically in the performed experiment for each of the 200 frame microscopic images of bone marrow smears. The ratio of the actual number of white blood cells to cells, which were found by software, characterizes the accuracy of the computer system. According to the study, at the stage of using an approach combining the method of the watershed transformation and distance without proposed improvements, the resulting accuracy was 56%, and after the application of our approach accuracy was 91% for white blood cells lined up (Figure 1) and 88% for leukocytes, forming a closed chain (Figure 4).

3. Conclusion

The method for segmentation of contacting leukocytes in microscopic images of bone marrow smears was proposed. The method is based on the analysis of structure of objects of a separation and distances filter in combination with the watershed method and distance transformation method. Application of the method improved the accuracy of the segmentation from 56 to 91%

Testing the proposed method on the extended sample of smears of bone marrow and using the proposed approach in automated systems of the analysis of smears of bone marrow will be further step of our research.

Acknowledgments

The reported study was funded by RFBR according to the research project № 17-07-01496.

IOP Conf. Series: Journal of Physics: Conf. Series **945** (2017) 012006 doi:10.1088/1742-6596/945/1/012006

References

[1] Mohapatra S *et al.* 2014 *Computer Methods in Biomechanics and Biomedical Engineering: Imaging & Visualization* pp 1-14
[2] Lewis S *et al.* 2012 *Dacie and Lewis practical haematology* (Amsterdam: Elsevier Health Sciences Publisher) p 600
[3] Reyes H *et al.* 2015 *20th Symposium on Signal Processing, Images and Computer Vision (STSIVA 2015)* (Bogota Colombia September 02 – 04 IEEE) pp 1-9
[4] Putzu L and Di Ruberto C 2013 *International Journal of Medical, Health, Biomedical, Bioengineering and Pharmaceutical Engineering* **7**(1) 20-27
[5] Da Costa L. 2015 *Clinics in laboratory medicine* **35**(1) 105-122
[6] Mishra S *et al. Biomedical Signal Processing and Control* **33** 272-280
[7] Gao Ch and Zhu H 2012 *Signal Processing* **92**(10) 2552–2559
[8] Acharjya P P *et al.* 2013 *International Journal of Innovative Research in Computer and Communication Engineering* **1**(2) 185-189
[9] Zakharenko Y V et al. 2017 *Journal of Physics: Conference Series* **798**(1) 012127
[10] Shankar V *et al.* 2016 *Advances in Computer Applications (ICACA) (IEEE International Conference on. IEEE* San Francisco USA 19-21 October) pp 186-189
[11] Amin M *et al.* 2015 *Journal of Medical Signals & Sensors* **5**(1) 49
[12] Nikitaev V G *et al.* 2015 *Proceedings of the 6th International Confererence on Bioscience and Bioinformatics (ICBB`15)* (Dubai United Emirates February 22-24 2015) 120-121
[13] Nikitaev V G *et al. 2015 Proceedings of the 15th International Conference on Signal Processing, Computational Geometry and Artificial Vision (ISCGAV`15)* (Sliema Malta August 17-19) p 49
[14] Nikitaev V G *et al. 2016 INTERNATIONAL JOURNAL OF BIOLOGY AND BIOMEDICAL ENGINEERING* **10** 109-114
[15] Nikitaev V G *et al.* 2014 *Measurement Techniques* **57**(5) 560
[16] Nikitaev V G *et al.* 2015 *WSEAS TRANSACTIONS on BIOLOGY and BIOMEDICINE* **12**(3) 16-19
[17] Nikitaev V G *et al.* 2015 *Measurement Techniques* **57**(10) 1203

IOP Conf. Series: Journal of Physics: Conf. Series **945** (2017) 012007 doi:10.1088/1742-6596/945/1/012007

Approach to building knowledge bases in information-measuring systems diagnostics of acute leukemias

V G Nikitaev, A N Pronichev, E V Polyakov, V V Dmitrieva

National Research Nuclear University MEPhI (Moscow Engineering Physics Institute), Kashirskoe shosse 31, 115409, Moscow, Russia.

E-mail: VGNikitayev@mephi.ru

Abstract. The paper describes an approach for the formation of the reference base of peripheral blood cells and bone marrow in information-measuring systems of acute leukemia diagnostics. The proposed approach has allowed to create a system, that is enable peer evaluation of blood cells needed for the training of recognition systems when carrying out microscopic studies.

1. Introduction

The current stage of development of measurement technology for diagnosis and condition monitoring of objects of different nature is characterized by an increasing complexity of tasks and intensive implementation of intelligent technologies in the measurement process. Expert systems are foundation of intelligent technologies. They are based on accumulating experience and knowledge of highly qualified experts.

One of the areas of works in the field of creation and development of expert systems that can accumulate the experience and expertise of highly skilled physicians is the diagnosis of acute leukemia.

The basic idea of these systems is to use the knowledge and experience of specialists of high qualification in the subject area by less skilled specialists, while addressing emerging problems.

The problem of the development of a knowledge model based on standard samples of micro specimens of blood and bone marrow remains unexplored area at present. Thus the knowledge base has special importance, if it is built on unique standard samples of large medical center in the end of expert measurements.

Knowledge base and database images are some of the key elements of building a modern decision support complex for medical diagnosis. Such complexes should include clinical system, research system, education system, information and telemedicine system on a single hardware and software platform[1].

Example of the use of software systems for the diagnosis of acute leukemia is intelligent system of decision support in the diagnosis of acute leukemia using microscopic images of blood cells presented in[2].

Evaluation of the effectiveness of a medical decision support system is given in [3]. Over 96% accuracy in the diagnosis of acute lymphoblastic leukemia was obtained for a sample of 180 images. Also method was proposed for allocation of several leukocytes in the field of view with an accuracy over 93%. The method was tested on 98 different sets. Diagnostic system was proposed for the detection of acute lymphoblastic leukemia (ALL) based on the analysis of shape and texture and two-dimensional classifier according to the method of support vectors [4]. In general the accuracy of the developed system is more 89% for the best combination of characteristics. A common disadvantage of these systems is the small volume of the experimental sample for confirming of the statistical significance of the results.

There is a need in the reference images for the development of methods and tools of medical decision-making systems and to verify the adequacy of the developed methods at the present stage of development of computer vision systems, particularly in the diagnosis of acute leukemia. The necessity of development of such systems is due to the lack of large digitized databases of images, lack of description of cells images[5-8].

The aim of this work is to develop an approach to form a base of knowledge in information-measuring systems of acute leukemia diagnostics.

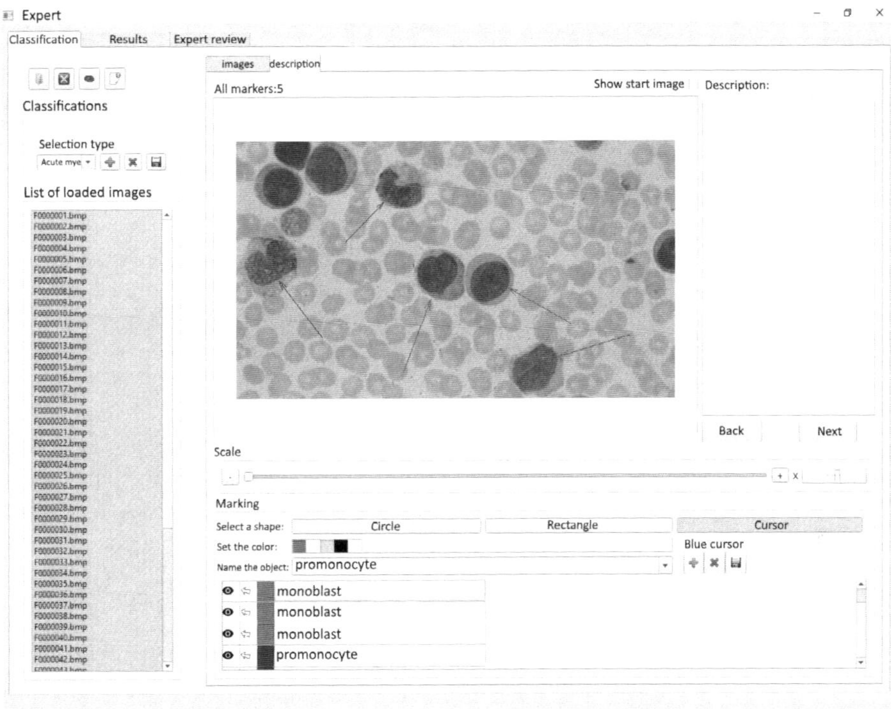

Figure 1. Example of the markup of images to generate the reference sample images.

2. Materials and methods

The evaluation of performance of approach was evaluated by doctors of clinical diagnostic laboratory of N. N. Blokhin Russian Cancer Science Center The images sample was formed from smears of peripheral blood of 10 donors and 6 patients with acute myeloid leukemia.

Images of lymphocytes and blasts were performed on stained preparations in the system of computer microscopy (Automated Olympus BX43 microscope with camera Imperx IPX-4M1ST-GCFB). Images were saved in BMP format, color-coded RGB24 (more than 16 million colors).

3. Results

Program was developed to generate the reference database which was built on the receipt of opinions from several experts and on the basis of collective opinion formation samples for research. Block diagram of the operation of the program is shown in Fig.2

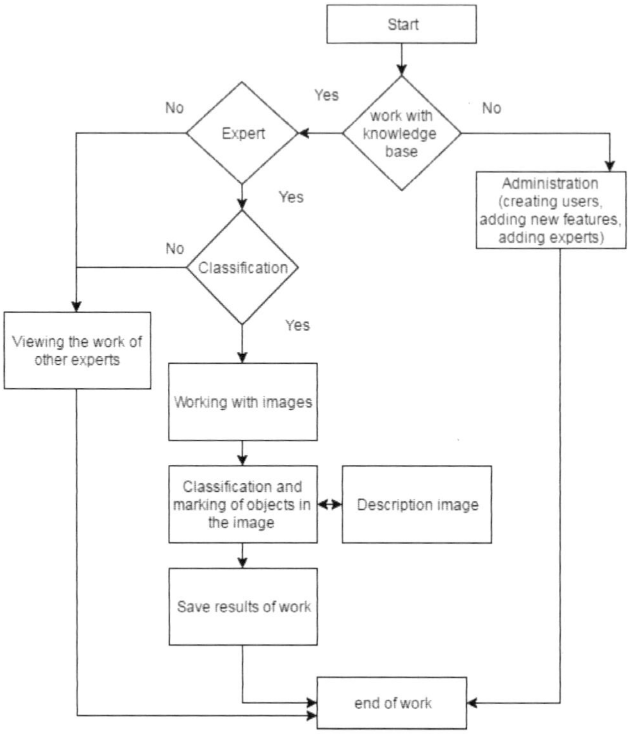

Figure 2. The block diagram of the program of knowledge building of the types of images in the diagnosis of acute leukemia.

The proposed solution is implemented with use of the library in Qt5 and C++ programming language. Interface for the markup of image obtained from the preparation is shown in Fig.1.

The applicability of the proposed approach was studied. 492 image frames were listed in the knowledge base. a few blood cells could be on one image frame. Bone morrow preparations of patients with acute myeloid leukemia variant M5Б were used. Cells in images were marked. There were 939 monoblast, 423 promonocyte, 27 lymphocytes, 12 segmented neutrophils, 6 myelocytes, 3 eosinophiles, 1 normoblast and 1 promyelotic.

4. Conclusion

The decision support system for the diagnosis of acute leukemias was developed. The importance of this system is determined by the necessity for the doctor-diagnostician to be able, in difficult cases, refer to the "reference view" - reference database. The developed system forms this opinion on base experts knowledge about the images of blood cells and bone marrow. The system allows the doctor to get an expert opinion of leading experts to determine the types of cells. In addition the system provides for computer recognition system developer the ability to generate samples for testing the developed system with simulating different levels of difficulty for the diagnosis of acute leukemia.

Acknowledgments

The reported study was funded by RFBR according to the research project № 17-07-01496

References

[1] Nikitaev V G 2015 *Measurement Techniques* **58**(6) 215

[2] Shivhare S and Shrivastava R 2012 *International Journal of Scientific & Technology Research* **1**(4) 125

[3] Madhukar M, Agaian S and Chronopoulos A T 2012 *Proceedings of SPIE - The International*

Society for Optical Engineering **8295**(829518) 12
[4] Rawat J *et al.* 2015 *Procedia Computer Science* **70** 748-756
[5] Neoh S C *et al.* 2015 *Scientific Reports* vol 5
[6] Nikitaev V G 2015 *Measurement Techniques* **58**(4) 68
[7] Nikitaev V G 2015 *Measurement Techniques* **58**(2) 215
[8] Zaytsev S M *et al.* 2017 *Journal of Physics: Conference Series*. **798**(1) 012133

IOP Conf. Series: Journal of Physics: Conf. Series **945** (2017) 012008 doi:10.1088/1742-6596/945/1/012008

Textural characteristics of bone marrow blast nucleus images with different variants of acute lymphoblastic leukemia

V G Nikitaev[1], A N Pronichev[1], E V Polyakov[1], A V Mozhenkova[2], N N Tupitsin[2], M A Frenkel[2]

[1]National Research Nuclear University MEPhI (Moscow Engineering Physics Institute), Kashirskoe shosse 31, 115409, Moscow, Russia.
[2]N.N. Blokhin Russian Cancer Research Center, Kashirskoe shosse 23, 115478, Moscow, Russia

E-mail: VGNikitayev@mephi.ru

Abstract. The paper describes the method of recognition of T - and B - variants of acute lymphoblastic leukemia in microscopic images of blood cells. The method is based on the use of texture characteristics of images. Experimental recognition accuracy evaluation is obtained from the sample of 38 patients (17 with T-ALL and 21 with B-ALL variants of acute lymphoblastic leukemia). The obtained results show the possibility of applying of the proposed approach to the differential diagnosis of T- and B- variants of acute lymphoblastic leukemia.

1. Introduction.

Diagnosis of acute lymphoblastic leukemia (ALL) and their variants is based on integrated morphocytochemical and immunophenotypic study of blasts in bone marrow aspirate [1, 2]. The study of cells in stained preparations includes evaluation of their various parameters: size, shape, and characteristics of the nuclei chromatin structure: thin netting, lumpy or dense. Blasts nuclei structure polymorphism is associated with a variety of immunophenotypical features, due to their belonging to different lines of hemopoiesis of B- or T-direction. Morphological heterogeneity of blast population is noted at T- and B- variants of ALL. Attempts of identifying of a logical relationship between the chromatin structure of blasts nuclei and immunophenotypic status has failed to date. Computer microscopy with using a multispectral camera has more wide opportunities to study the structure of nuclear chromatin fibres than optical microscopy. It allows you to objectify the data obtained in the form of numerical indices. Digital expression of the results of the study provides detailed analysis of the peculiarities of the blasts nuclei structure images in preparations of bone marrow of patients with B- and T- ALL [3-8].

The aim of this work is the comparative evaluation of texture characteristics of nuclei chromatin of bone marrow blasts with T- and B- variants of ALL. Wavelet transformation and spatial adjacency matrix were used for texture characteristics calculations.

2. Materials and methods

The sample of preparations of 38 patients was formed. 17 of them had T-ALL and 21 patients had B-ALL. The diagnosis was revealed on using morphocytochemical and immunophenotypic studies, which were carried out in the laboratory of immunology of hematopoiesis of N. N. Blokhin Russian Cancer Science Center (head of the laboratory MD. Professor N. N. Tupitsyn).

The automated microscope Olympus BX43 with the Imperx camera IPX-4M1ST-GCFB was used for shooting digital images.

4930 images of bone marrow lymphoblasts were obtained for the study. There were 2048 lymphoblasts images from bone marrow preparations of patients with T-ALL and 2882 lymphoblasts images from bone marrow preparations of patients with B-ALL. Textural and wavelet features of blasts nuclei were calculated for the obtained images. Color model RGB, XYZ, HSL, Lab, Luv, LCH, HLS, HSV, YUV, YIQ, YCbCr, CMY were used for computation [9-13]. The areas with a size of 45x45 pixels inside blasts nuclei were selected for analyze. 513 types of textural characteristics for each of the cells were calculated. Then the characteristics averages for the cells sample in the preparations of each of patients were calculated.

The ALL type recognition was carried out using a linear classifier with a distance function Euclidean, Manhattan and Chebyshev in two-dimensional feature space [8]. We used all possible pairs of averages for all 513 textural characteristics types.

Table 1. Generalized characteristics M_{ENT} and M_{RENGE}^{-1} for the group of T - and B-cells of ALL patients

The patient (type of blasts)	Quantity of cells	M_{ENT}	M_{RENGE}^{-1}	The patient (type of blasts)	Quantity of cells	M_{ENT}	M_{RENGE}^{-1}
P 1(B)	159	5,778	72,501	P 20(B)	181	5,824	68,640
P 2(B)	176	5,373	66,157	P 21(B)	193	6,104	69,871
P 3(B)	166	5,798	71,040	P 22(T)	82	5,904	78,917
P 4(B)	103	5,664	69,700	P 23(T)	154	5,527	69,256
P 5(B)	100	5,874	72,396	P 24(T)	100	5,574	70,021
P 6(B)	183	5,887	73,818	P 25(T)	121	5,784	73,764
P 7(B)	146	5,730	66,776	P 26(T)	83	5,791	74,632
P 8(B)	170	6,055	72,907	P 27(T)	130	5,740	75,390
P 9(B)	117	6,120	70,409	P 28(T)	131	6,176	77,764
P 10(B)	116	5,898	72,213	P 29(T)	124	5,518	75,200
P 11(B)	107	5,640	68,775	P 30(T)	172	5,683	75,799
P 12(B)	33	5,867	77,986	P 31(T)	76	5,895	75,241
P 13(B)	168	6,097	71,219	P 32(T)	37	5,619	73,221
P 14(B)	51	5,921	71,565	P 33(T)	153	5,357	70,243
P 15(B)	115	5,954	74,195	P 34(T)	181	5,701	80,181
P 16(B)	73	6,007	73,637	P 35(T)	113	5,542	71,888
P 17(B)	185	5,798	65,043	P 36(T)	124	5,701	73,003
P 18(B)	214	5,607	66,611	P 37(T)	128	5,656	71,110
P 19(B)	126	5,637	67,796	P 38(T)	139	5,274	67,594

3. Results

M_{ENT} and M_{RENGE}^{-1} were identified as a couple of signs, in which the ALL variants detection occurred with minimal error. M represents the average value for the blast cells of the patient preparation. Index "ENT" corresponds to the texture characteristic entropy, calculated for R (red) coordinate in the RGB color space. The index "$RENGE^{-1}$" corresponds to the value of wavelet characteristic "relative range" for U (color-difference) component in the YUV color space. The data obtained are presented in Table1. A graphical representation of Table1 is shown in figure 1.

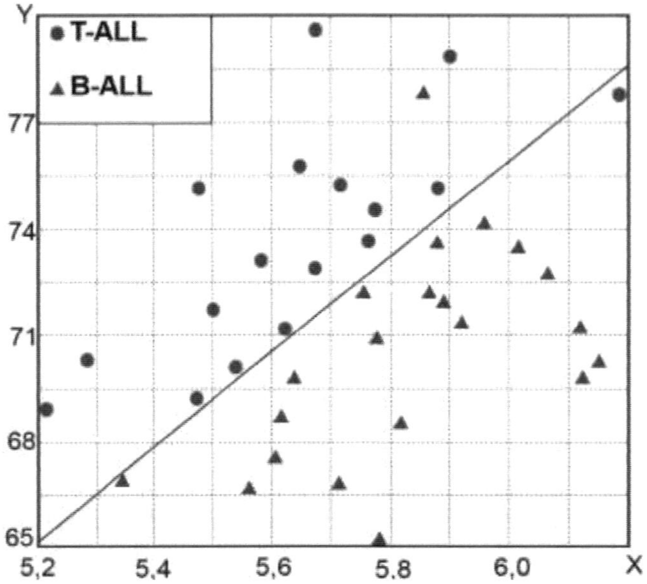

Figure 1. The distribution of the two classes (T-ALL and B-ALL) in two-dimensional feature space. M_{ENT} feature is marked on the x-axis, and $M_{RENGE}{}^{-1}$ is marked on the y - axis.

Thus, on these characteristics, you can judge the ALL variant (T - or B - type). Using of these characteristics allowed to achieve 95% recognition accuracy of ALL variants (T - or B – type).

Generalized criterion C_{GC} was proposed for the practical evaluation of the ALL type. The value of this criterion is calculated according to the formula: $C_{GC} = M_{ENT} \cdot 0{,}998 + M_{RENGE^{-1}} \cdot (-0{,}065)$.

C_{GC} criterion distribution histogram for two classes of patients with ALL T-type and B-type is presented in figure 2.

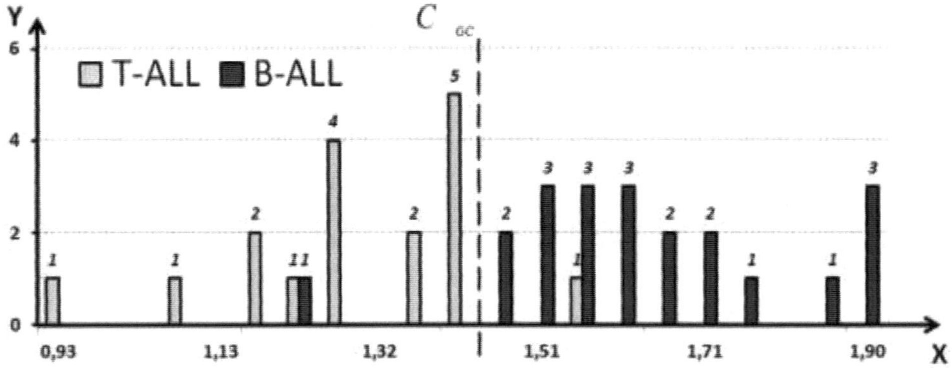

Figure 2. C_{GC} criterion distribution histogram for two classes of patients with ALL T-type and B-type.

Using a criterion value C_{GC} = 1.405, you can diagnose variant of acute lymphoblastic leukemia according to the rule: if $C_{GC} > 1.405$, then B-ALL type is, if $C_{GC} \leq 1{,}405$, then T-ALL type is [14].

Thus, a study based on computer analysis of images of the nuclear chromatin structure showed effective application of the apparatus of the texture analysis for establishing the T- and B- variants of acute lymphoblastic leukemia.

4. Conclusion

The study showed that the use of the texture analysis method of the blood cells nuclei images on microscopic examination of bone marrow aspirates allows to distinguish T- and B-variants of acute lymphoblastic leukemia. In experimental studies were used bone marrow aspirates preparations from 17 patients with T-ALL and 21 patients with B-ALL.

In further experiments we plan to increase the volume of the preparations sample. In addition, research will be conducted to establish correlations between morphological and immunophenotypical characteristics of lymphoblasts L1, L2, L3 types, pre-B, pre-pre-B, Pro-B and pre-T types for the differential diagnosis of appropriate modifications of acute leukemia with the use of computer microscopy.

Acknowledgments
The reported study was funded by RFBR according to the research project № 17-07-01496

References

[1] Frenkel M A 2007 *Laboratory diagnosis of acute leukemia Clinical Hematology* edited by Professor M A Volkova (original Russian title: *Laboratornaya diagnostika ostryh leykozov Klinicheskaya onkogematologiya* pod redaktsiey prof. M.A.Volkovoy) (Moscow Medicine publisher) pp 306-319
[2] Tupitsin N N 2007 *Immunodiagnostics of acute leukemia and non-Hodgkin's lymphoma. Clinical Hematology* edited by Professor M. A. Volkova (original Russian title: *Immunodiagnostika ostryh leykozov i nehodzhkinskih limfom. Klinicheskaya onkogematologiya* pod redaktsiey prof. M.A.Volkovoy) (Moscow Medicine publisher) pp 338-370
[3] Frenkel M A 2014 (in Russia) *Haematopoiesis immunology* **12**(1-2) 18
[4] Volkova S A and Borovkov N N 2013 *Fundamentals of clinical Hematology* (original Russian title: *Osnovi klinicheskoy gamatologii*) (N. Novgorod: Lower GMA publisher) p 400
[5] Fiyas A T 2013 (in Russian) *Hematological malignancies: clinical features, diagnosis, treatment* (original Russian title: *Gemoblastosi: klinika, diagnostika, lechenie*) (Grodno: GrGMU publisher) p 172
[6] Swerdlow S H et al. 2008 *WHO classification of tumours of haematopoietic and lymphoid tissues*, (IARC: Lyon)
[7] Shivhare S and Shrivastava R 2012 *International Journal of Scientific & Technology Research* **1**(4) 3 125
[8] Nikitaev V G et al. 2015 *Measurement Techniques* **57**(10) 1203
[9] Pronichev A. N et al. 2017 *Journal of Physics: Conference Series* **784**(1) 012003
[10] Nikitaev V G et al. 2015 *TRANSACTIONS on BIOLOGY and BIOMEDICINE* **12**(3) 16-19
[11] Nikitaev V G et al. 2015 *Proceedings of the 6th International Confererence on Bioscience and Bioinformatics (ICBB`15)* (Dubai United Emirates February 22-24) p 120
[12] Nikitaev V G et al. 2015 *Proceedings of the 15th International Conference on Signal Processing, Computational Geometry and Artificial Vision (ISCGAV`15)* (Sliema Malta August 17-19) p 49
[13] Nikitaev V G et al. 2016 *International journal of biology and biomedical* engineering 10 2016 p 109
[14] Nikitaev V G et al. 2016 *Bulletin of the Lebedev Physics Institute* **43**(10) 306-308

IOP Conf. Series: Journal of Physics: Conf. Series **945** (2017) 012009 doi:10.1088/1742-6596/945/1/012009

Sapphire implant based neuro-complex for deep-lying brain tumors phototheranostics

A S Sharova[1,2], YU S Maklygina[2], G M Yusubalieva[3], I A Shikunova[4], V N Kurlov[4] and V B Loschenov[1, 2]

[1] National Research Nuclear University MEPhI, Moscow, Russia
[2] Prokhorov General Physics Institute, Russian Academy of Sciences, Moscow, Russia
[3]Serbsky National Research Center for Social and Forensic Psychiatry, Moscow, Russia
[4] Institute of Solid State Physics, Russian Academy of Sciences, Chernogolovka, Russia

E-mail: alina.s.sharova@gmail.com

Abstract. The neuro-complex as a combination of sapphire implant optical port and osteoplastic biomaterial "Collapan" as an Aluminum phthalocyanine nanoform photosensitizer (PS) depot was developed within the framework of this study. The main goals of such neuro-complex are to provide direct access of laser radiation to the brain tissue depth and to transfer PS directly to the pathological tissue location that will allow multiple optical phototheranostics of the deep-lying tumor region without repeated surgical intervention. The developed complex spectral-optical properties research was carried out by photodiagnostics method using the model sample: a brain tissue phantom. The optical transparency of sapphire implant allows obtaining a fluorescent signal with high accuracy, comparable to direct measurement "in contact" with the tissue.

1. Introduction

Currently, treatment of deep-lying brain neoplasms is characterized by a complex approach, including surgical removal of the tumor with following appropriate therapy [1-2]. Due to the characteristic features of such neoplasm type (high metastasis grade, deep invasiveness and rapid proliferation), the possibility of radical tumor resection is not surgically available. Thus, to reduce the likelihood of relapses, the remained after the surgical intervention tumor cells should be therapeutically treated. Photodynamic therapy (PDT) and photodiagnostics (PD) are considered to be the most effective concomitant methods of oncological diseases theranostics [3-5]. However, due to the deep localization of brain tumors, multiple PD and PDT are limited and require repeated surgical intervention. Moreover, the presence of a blood-brain barrier prevents the penetration of drugs into the brain tissue that also limits subsequent therapy. In this way, the development of neuro-complex, providing permanent access to the bed of deep-lying tumor during postoperative period, seems to be promising. Such kind of phototheranostics neuro-complex will provide local delivering of drug and laser radiation to the neoplasms bed. In the future, this development will make it possible to translate an incurable disease into a chronic one by permanent monitoring of neoplasm processes.

IOP Conf. Series: Journal of Physics: Conf. Series **945** (2017) 012009 doi:10.1088/1742-6596/945/1/012009

The purpose of this research is to study the spectral-optical properties of the developed neuro-complex on the brain tissue model by PD method.

2. Materials and methods

Optimally selected sapphire implant construction underlies the developed neuro-complex. Sapphire is a mechanically strong material, has a high inertia to biological tissues including brain tissues [6, 7], and characterized by a wide range of wavelengths transmission [8]. Sapphire bone implants show prolonged osseointegration. Sapphire implant has an optimal geometry with a cylindrical foundation (D_{out} = 3 мм, h = 5 мм) for deep-lying brain tissues research. It also has an expansion of one of the ends for the skull fixation. Implant was obtained by growing sapphire capillaries according to the EFG / Stepanov method [9] with minimal finishing machining (Figure 1a).

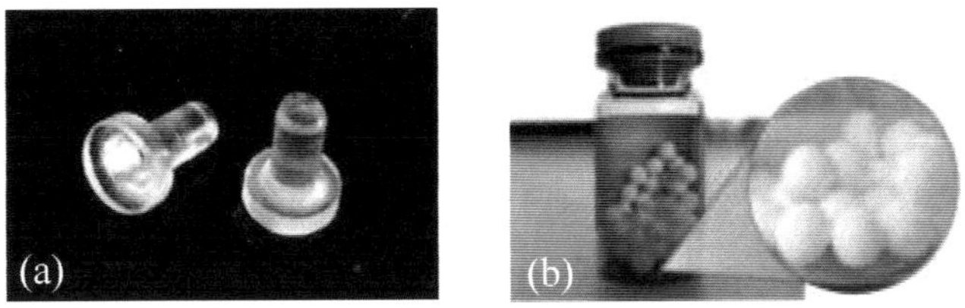

Figure 1. a) Sapphire implants as an optical port; b) Osteoplastic biomaterial "Collapan"*.

Aluminum phthalocyanine nanoparticles (nAlPc) were used as a drug (PS) for phototheranostics (synthesized by Organic Intermediates & Dyes Institute (NIOPIK), Russia). The osteoplastic biomaterial "Collapan" (Figure 1b), which includes 55% of collagen and 45% of hydroxyapatite (produced by LLC "Intermedapatit"*) acted as a PS carrier. The deposition of nanoparticles on the internal and surface structures of biomaterial was carried out by incubation of «Collapan» biomaterial fragment in the aqueous nAlPc colloid (c = 1 mg / ml). With the purpose of the optical neuro-complex approbation, the model reproduced the optical and structural properties of the brain tissue (phantom) was created. For brain tissue phantom preparation the scattering and absorbing mediums are necessary. 20% intralipid was used as a scatterer and distilled water acted as an absorber. The phantom of following composition: 1.1% intralipid in aqueous solution repeats optical properties of the brain tissue with a high accuracy. The developed neuro-complex optical properties research was carried out by photodiagnostics using the fiber-optic spectrometer LESA-01 "BIOSPEC" (Russia). The fluorescence was excited by a laser source at a power density of ~ 100 mW / cm^2 and λ = 632.8 nm wavelength, chosen in accordance with the nAlPc absorption maximum.

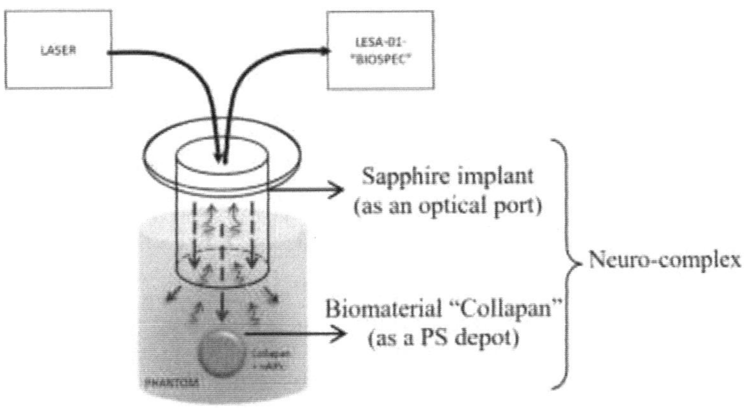

Figure 2. Scheme of the experiment in a photodiagnostics mode. (Sapphire implant in a brain tissue phantom with osteoplastic "Collapan" material as a carrier of nAlPc PS)

The developed neuro-complex was also tested in vivo on laboratory animals: mature rats with induced C6 glioma. After resection of the larger central tumor part, the neuro-complex was placed in the tumor bed and fixed at the skull base in the striatum region. MRI-control of developed optical system biocompatibility was carried out by 7 T-biospectrotomograph BioSpec 70/30 USR (Bruker). The optical part of neuro-complex research is still in progress and carried out currently, namely, the photodiagnostics efficiency of implantation field processes and non-invasive phototherapy of pathological hotbed by means of developed construction.

3. Results and discussion

The spectral optical properties of both individual and groups of optical neuro-complex components were investigated during the study. It was found that sapphire implant, "Collapan" biomaterial and brain tissue phantom do not show fluorescent properties on their own (Figure 3). Free nAlPc nanocrystals in aqueous colloid, not capable initially to luminescence, are photoactivated by interaction with biomaterial structure and the model medium of brain tissue. During fluorescent signal obtaining through sapphire implant, the slight decrease of intensity (7%) due to the optical losses in the material was registered. Thus, the developed neuro-complex based on sapphire implant contributes obtaining a fluorescent signal with high accuracy, comparable to direct measurement "in contact" with tissue (Figure 2). The process of nAlPc PS carrier "Collapan" biointegration was simulated by exposure of Dimethyl sulfoxide polar solvent (DMSO) on the sample surface, that simulated the in vivo conditions process of PS carrier interaction with biocomponents. The observed intensity increase of PS fluorescence signal evidences the nanoparticles photoactivation. It was also found that PS nanoparticles are strongly connected with the "Collapan" structure, since even by aqueous dispersion the washout of nanocrystals from the material was not achieved. The dissolution-washout cycle of nAlPc was repeated multiple times, the spectra are presented for two consecutive exposures (Figure 3b).

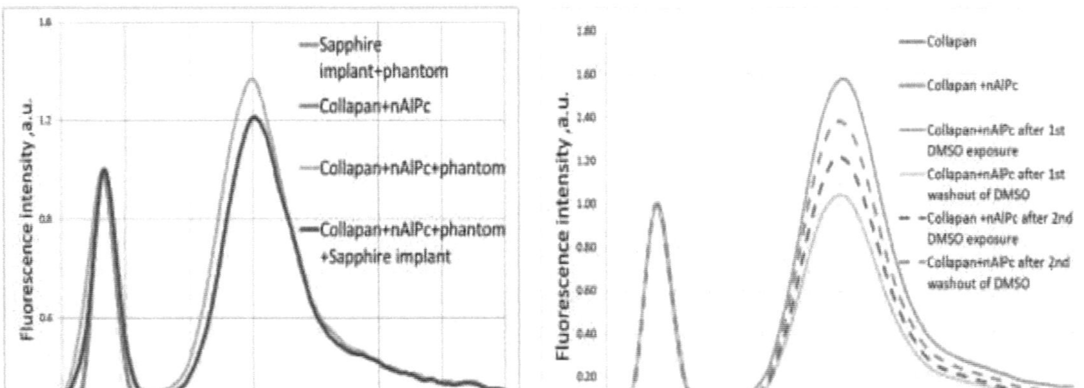

Figure 3. a) The fluorescence spectra obtained for individual and groups of optical neuro- complex components; b) The fluorescence spectra of "Collapan" osteoplastic material as a carrier of nAlPc PS under different conditions (Two dissolution-washout cycles of nAlPc are presented).

For optimal physical characteristics and biocompatibility issues of complex research, an MRI-monitoring of laboratory animals with an induced brain tumor was conducted during the postoperative period. MRI images of laboratory rats cerebral skull part were obtained, indicating the absence of inflammatory reactions in the implantation area (Figure 4). The implant rejection cases were not also revealed. That circumstance allows considering the chemical and physical characteristics of this complex to be optimal and proves the prospects of further research.

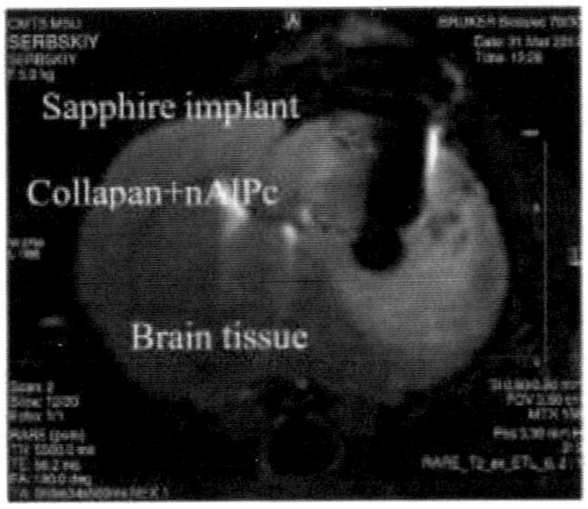

Figure 4. MRI image of the neuro-complex (sapphire implant with osteoplastic "Collapan" material as a carrier of nAlPc PS) in the brain tissue of a laboratory rat with induced C6 glioma.

4. Conclusion

The spectral optical properties of the developed neuro-complex: sapphire implant as an optical port + osteoplastic "Collapan" biomaterial as a carrier of nAlPc PS nanoparticles were investigated in the framework of this study. Sapphire implant allows obtaining a fluorescent signal with high accuracy, comparable to direct measurement "in contact" with tissue. "Collapan" biomaterial seems to be a promising PS carrier that is confirmed by resistance of nanocrystals to washout in time. The results of laboratory animals with brain glioma MRI analysis showed no adverse reactions to the developed neuro-complex. In this way, in order to carry out multiple optical deep-lying tumor phototheranostics without repeated surgical intervention, further studies of the developed neuro-complex in vivo seem to be promising.

Acknowledgments

This work was supported by Competitiveness Program of National Research Nuclear University MEPhI.

References

[1] Wiesner S M, Freese A and Ohlfest J R. 2005 Emerging concepts in glioma biology: implications for clinical protocols and rational treatment strategies *Neurosurg.* **19** E3 .

[2] Wang G, Fu XL, Wang JJ, Guan R and Tang XJ 2016 Novel strategies to discover effective drug targets in metabolic and immune therapy for glioblastoma *Curr. Cancer Drug Targ,* **17** 17-39.

[3] Brendan J Quirk, Garth Brandal, Steven Donlon, Juan Carlos Vera, Thomas S Mang, Andrew B Foy, Sean M Lew Albert W Girotti , Sachin Jogal, Peter S LaViolette, Jennifer M. Connelly and Harry T Whelan 2015 Photodynamic therapy (PDT) for malignant brain tumors — Where do we stand? *Photodiagnosis and Photodynamic Therapy* **12** 530—544

[4] Potapov A A, Nazarov V V, Goryaynov S A, Okhlopkov VA, Shishkina L V, Shurkhay V A, Loschenov V B, Saveleva T A, Kuzmin S G and Chumakova A P 2014 A case of brain abscess mimicking cystic brain tumor and showing intraoperative 5-aminolevulinic acid fluorescence (case report) *J. Chirurgia* **27** 257-260.

[5] Jiro Akimoto 2016 Photodynamic Therapy for Malignant Brain Tumors *Neurol. Med. Chir. (Tokyo)* **56** 151–7

[6] Shikunova I A and Kurlov V N 2016 Sapphire shaped crystals for medicine *Journal of Physics: Conference Series* **672** 012018

[7] Khan S P, Auner G G, Palyvoda O et al. 2011 Biocompatibility assessment of next generation materials for brain implantable microelectrodes *Materials Letters* **65** 876–9.

[8] Kurlov V N 2016 Sapphire: Properties, Growth, and Applications *Reference Module in Materials Science and Materials Engineering* pp 1-11

[9] Kurlov V N, Rossolenko S N, Abrosimov N V and Lebbou Kh 2010 Shaped Crystal Growth *Crystal Growth Processes Based on Capillarity* chapter 5 pp 277-354

The method for glomerulations detection in histological images of prostate

A A Zavarzin[1], A N Pronichev[1], O V Rodionova[1], E A Komochkina[1], E A Prilepskaya[2], M V Kovylina[2]

[1] National Research Nuclear University MEPhI (Moscow Engineering Physics Institute), Kashirskoe shosse 31, 115409, Moscow, Russia
[2] Moscow State University of Medicine and Dentistry named after A.I. Evdokimov, Delegatskaya street, 20/1, 127473, Moscow, Russia

e-mail: Zavarzin.mephi@gmail.com

Abstract. In the work presented, a method for detecting glomeruli in pictures of histological preparations of the prostate gland is described, the presence of which indicates a malignant neoplasm. Pathological structures at the level of microimages are investigated. The developed method is the result of joint activity of the National Research Nuclear University "MEPhI" and the Moscow State Medical and Stomatological University named after A.I. Evdokimova.

1. Introduction

In the global structure of oncological morbidity, prostate cancer ranks sixth, and among men - the third. Today's crucial challenges are the development and implementation of automated software systems to diagnose prostate cancer. Solutions to these challenges could allow us to automate the diagnosis of cancer, increase its accuracy and, therefore, accelerate a decision-making process regarding the methods of medical intervention. This pathology is annually diagnosed in more than half a million people, or about one tenth of all oncological diseases in men [1]. At present, there is no counterpart to the developed method for detecting glomeruloid structures.

The task of the research stage is to extract and describe characteristic features of the pathological structures, called glomeruli, and distinguish them from other objects of the histological specimens. Figure 1 presents examples of glomeruli.

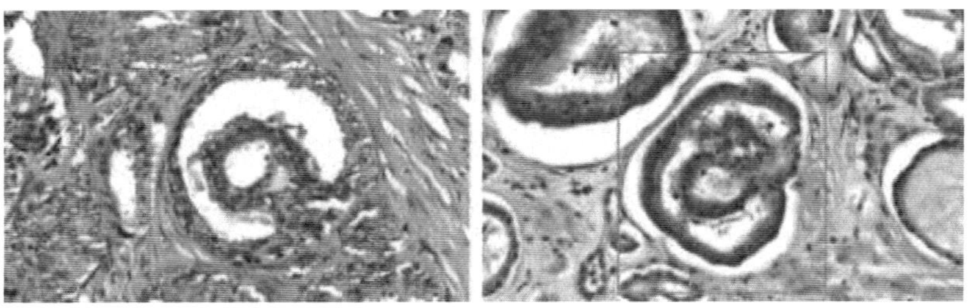

Figure 1. Histological images of glomeruli

IOP Conf. Series: Journal of Physics: Conf. Series **945** (2017) 012010 doi:10.1088/1742-6596/945/1/012010

The formation of glomeruloid bodies is a rare type of cancer differentiation grades. They are made up of solidifying malignant cells, surrounded by crescent-shaped spaces, and then by the wall of an enlarged acinus or duct with a lining of the cancerous epithelium. In accordance with the Gleason index, a tumor of such a differentiation generally corresponds to grade 4. On the contrary, healthy cells produce glandular formations containing spaces of round, ellipsoidal shape. An example of benign glandular structures is shown in Figure 2

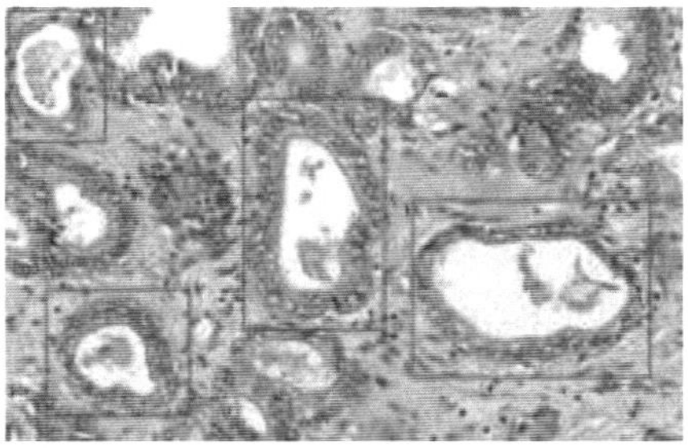

Figure 2. Histological images of benign glandular structures

The purpose of the work is to develop and automate the procedure for detecting glomeruloid structures in the images of histological specimens of the prostate gland.

2. Materials and methods

Histological specimens of the prostate gland are subjected to detailed analysis. The average size of such images is 1300 x 900 pixels; the color depth is 24 bits.

A binarization method is used to extract ducts of glandular structures (white spaces) contained both in the areas of benign cellular formations and in glomeruli. In the binarization process, the original halftone image having a certain number of brightness levels is converted to a black and white image whose pixels have only two values– 0 and 1 [2]. Threshold processing of the image can be carried out in different ways. To use a binarization method, we convert the original three-channel (RGB) image into a halftone image with a color depth of 8 bits (256 semitones). Further, we apply the binarization method with an upper threshold:

$$f'(m,n) = \begin{cases} 0, f(m,n) \geq t \\ 1, f(m,n) < t \end{cases}$$

Where $f(m,n)$ is the pixel half-tone level with coordinates (m,n), $f'(m,n)$ is the value (0/1) of the pixel as a result of binarization, t is the threshold value. Due to the study of histological specimens, the threshold value t can be determined.

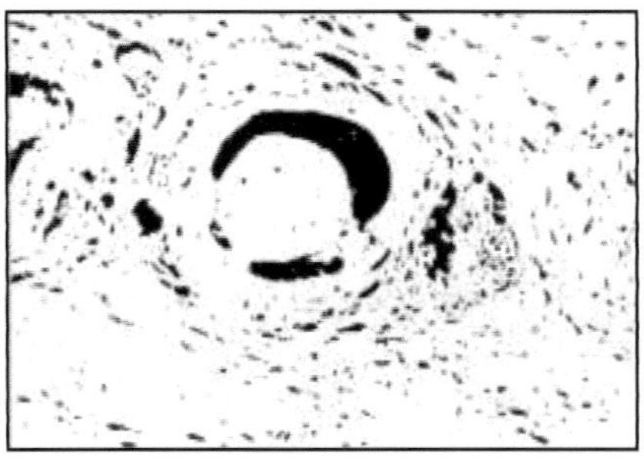

Figure 3. Binarization result

Figure 3 shows the outcome of applying the binarization method. The black channels are the ducts of the prostate. Further we extract the contours of the gaps for subsequent analysis of their shapes. In order to operate with contours, we present them in the form of a chain code of Freeman (Freeman Chain Code) [3]. Figure 4 shows the result of the contour extraction operation.

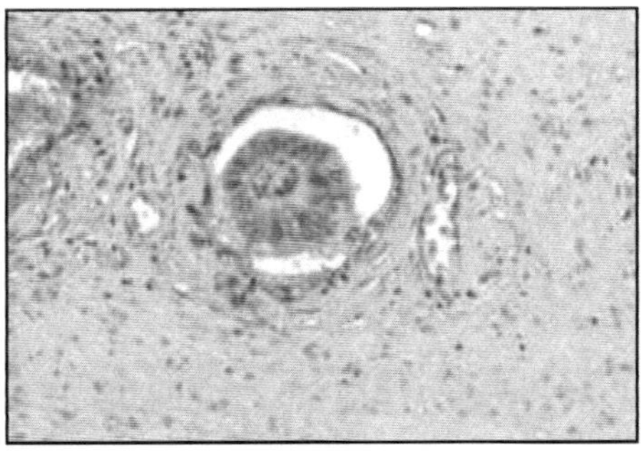

Figure 4. Result of the contour extraction

Then we extract crescent-shaped contours. Within the framework of the task set, the criterion for detecting such contours is as follows: if the center of mass of a contour does not belong to its inner area, then the contour is considered to be of crescent shape. With the purpose of determining whether or not the point belongs to the contour- bounded area, the ray tracing method can be used. Assume that it is necessary to define whether point A is related to K contour. To do this, we draw a straight line to point A from some remote point. Along that direction there can occur zero or several intersections of the boundary of the contour: we enter the contour at the first intersection, we exit from it at the second one .If we reach point A with an odd number of intersections of the contour boundary, then point A lies inside the contour, and if an even number of intersections is produced, then the point is outside the contour K. In this case, point A is the contour's center of mass. As shown in Figure 5, straight lines to point A are drawn along four directions.

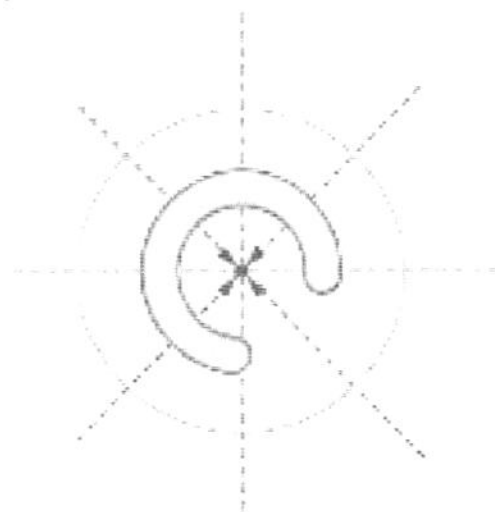

Figure 5. Determination of the point belonging to the contour area Direction of straight lines to the studied point

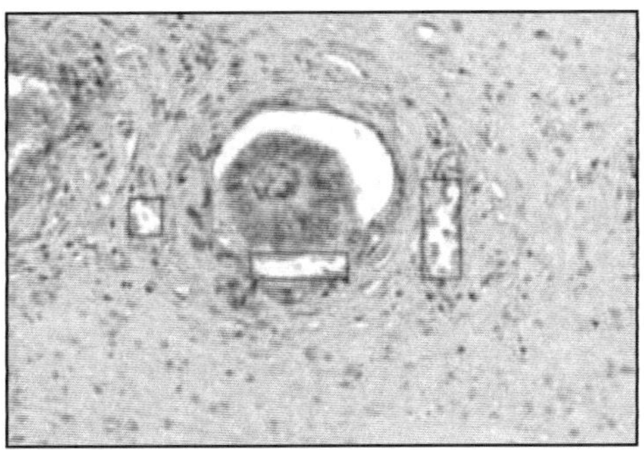

Figure 6. Result of extraction of crescent- shaped contours

Figure 6 shows the result of extracting crescent-shaped contours. The green rectangle contains a contour satisfying the above criterion, the blue ones - the rest contours.

3. Results

For computer image processing in C ++ using the Qt library and the OpenCV computer vision library [4], a software module with a minimal graphical user interface has been developed that allows downloading images and applying automatic analysis, as well as displaying results (Figure 7).

Precision and recall are the metrics used in evaluating the majority of the information extraction algorithms. The system precision within the class is the proportion of documents actually belonging to a given class relative to all documents that has assigned to this class by the system. Recall is the proportion of the classifier's documents belonging to the class relative to all documents of this class in the test set.

Three-channel images of histological specimens of the prostate with a size of 1,300 x 900 pixels with a color depth of 24 bits (8 bits per color component) were used as a test set. These images have been produced as a result of splitting the original image size of 60,000 x 130,000 pixels into tiles. The number of images used for the test set is 40. As a result of testing, the following values for precision and recall have been obtained:

$$\text{Precision} \approx 90\ \%,$$
$$\text{Recall} \approx 70\%.$$

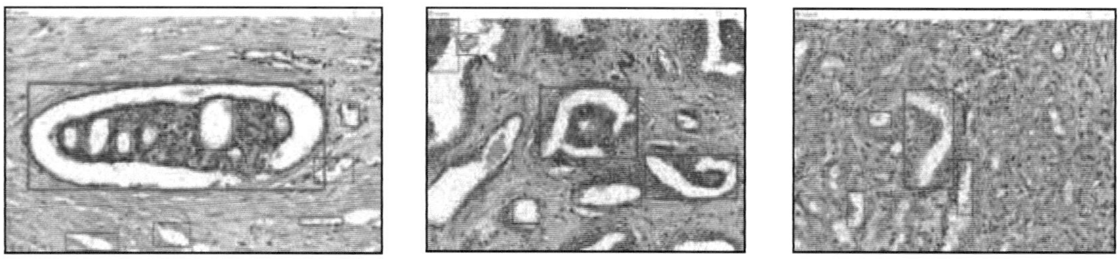

Figure 7. Algorithm results

Acknowledgments

This work was supported by Competitiveness Program of National Research Nuclear University MEPhI.

References

[1] Stratsky A. V., Yushko E. I. 2015 *Cancer of the prostate* (Minsk, BSMU) 16 pages [in Russian]
[2] Shapiro L. 2006 *Computer vision* (Moscow, Binom) 752 pages [in Russian]
[3] Freeman H. 1961 *On the encoding of arbitrary geometric configurations, IRE Transactions on Electronic Computers EC* **10** 260-268
[4] Bradski G., Kaehler A. 2008 *Learning OpenCV: Computer vision with the OpenCV library* (O'Reilly Media, Inc.) 556 pages

IOP Conf. Series: Journal of Physics: Conf. Series **945** (2017) 012011 doi:10.1088/1742-6596/945/1/012011

Application of silver films with different roughness parameter for septic human serum albumin detection by Surface Enhanced Raman Spectroscopy

A Y Zyubin[1], E I Konstantinova[1,2], K I Matveeva[1], V A Slezhkin[1,3], I G Samusev[1], M V Demin[1] and V V Bryukhanov[1]

[1] Science and Educational Centre "Fundamental and Applied Photonics. Nanophotonics", Immanuel Kant Baltic Federal University, 14 A, Nevskogo street,

236016 Kaliningrad, Russia

[2] Department of Physics, Kaliningrad State Technical University, 1 Sovetsky Ave.,

236022, Kaliningrad, Russia

[3] Department of Chemistry, Kaliningrad State Technical University, 1 Sovetsky Ave.,

236022, Kaliningrad, Russia

Email: azubin@mail.ru

Abstract. In this paper, the rough silver films parameters investigation, used as media for surface enhancement Raman spectroscopy for health and septic human serum albumin (HSA) study results have been presented. The detection of small concentrations of HSA isolated from blood serum and it main vibrational groups identification has been done.

1. Introduction

Raman spectroscopy is widely used in biomedical research [1], including non-invasive [2] and label-free [3] biomolecules diagnostics. Surface Enhanced Raman Spectroscopy (SERS) is used to identify both single biomolecules and the processes occurring in them [4].

SERS is explained by the effectiveness of the analysis of trace amounts of substances, which is claimed in various fields, such as biology, medicine, criminalistics, ecology, etc. The main advantages of the method include the simplicity of sample preparation, high sensitivity, ability to detect ultra-low concentrations of substances under study up to single molecules, molecules qualitative determination and detailed information on their structure and orientation by characteristic spectra obtaining [5-7].

This paper performs the investigation of silver films roughness parameters for both septic and health HSA SERS study. The main structural groups of the septic human albumin in the middle-band region with a frequency shift of 500-1700 cm^{-1} have been analyzed and compared with its healthy condition. It has been shown that surface plasmons (SPs) appear on a rough silver film when interacting with the optical radiation and multiply ($\sim 10^1 - 10^2$) Raman scattering signal for an albumin biomolecule.

2. Materials and Methods

The human plasma was thawed for 30 minutes to room temperature before experiment, then, a weak solution of 5-10% aqueous sulfasalicylic acid solution ($C_7H_6O_6S$) was prepared. Extraction of HSA was prepared by placing 2 ml of unfrozen plasma in an Eppendorf tube, then the solution of

IOP Conf. Series: Journal of Physics: Conf. Series **945** (2017) 012011 doi:10.1088/1742-6596/945/1/012011

sulfosalicylic acid was added to plasma dropwise, as a result, the hydrate coat of the protein was destroyed and its solubility in water decreased, and HSA precipitated in the bottom of the tube. Special attention was paid to a soft change process of the pH solution, both for the gradual withdrawal of albumin into the sediment, and for the subsequent renativation of a part of the protein. The supernatant was neatly removed with a pipette, the pH of the protein was adjusted to 5.5 (near the isoelectric point of the protein) by successive washing with distilled water and, as a result, adding up to 5 ml in the tube. Thus, the denaturing agent was partially removed and part of the HSA was transferred to its original state.

SERS substrates were prepared as follows: copper substrates were stripped and polished from one side, to gain a metallic luster. After polishing, the surface was washed with 96% ethanol, after it was degreased in a 5% solution of sodium hydroxide, washed with distilled water, and dried at 60° C. After drying and washing procedures, silver electrodeposition was deposited at 20-22° C temperature with current density of 5 mA/cm^2 for 15 minutes. Thickness of the silver film was measured to be 5 µm. Then, the roughness and porosity of one part of the sample surface was changed by anodic dissolution of the silver film onto 0.25; 0.5; 0.75 µm-thick layers at current density of 5 mA/cm^2. After electrodeposition and anodic dissolution, the samples were washed for 10 minutes with running water to remove the electrolyte salts and dried. The surface roughness was controlled by method described in [8] by SPM Certus Standart module with optical registration system (LTD «NanoScanTechnology», Russia) and tapping mode atomic force microscopy (AFM) use.

Raman spectra were obtained by CENTAUR U unit with diode-pumped solid-state laser (DPSS) excitation (λ = 633 nm, radiation power of 37 mW). Raman spectra from SERS surface with a drop of protein solution on the substrate at 11 mg/ml concentration were recorded using a CCD detector matrix with 120 s signal accumulation.

3. Results and Discussion

Figure 1 shows AFM image of SERS substrate with different anodic dissolution of the silver film layer (Figure 1A-1C) and SERS substrate with protein solution of high concentration (Figure 1D).

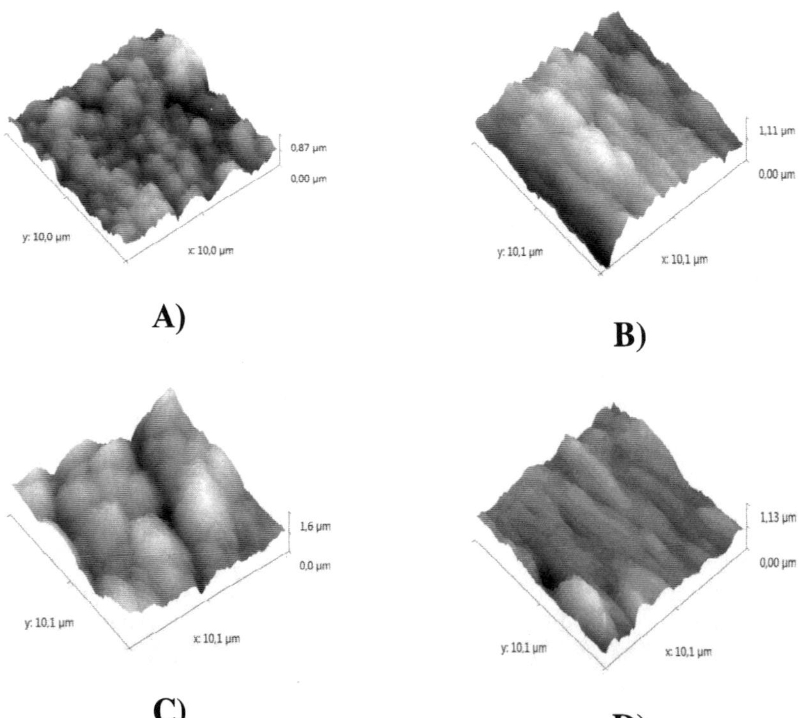

Figure 1. The morphology of the porous silver film surface, controlled by AFM for 0.75 μm (a), 0.5 μm (b), 0.25 μm (c) anodic dissolution of the silver film layer and for 0.5 μm and anodic dissolution of the silver film layer with 45 mg/ml protein solution on the surface (d)

The obtained results showed that increase in the anodic dissolution value from 0.25 μm to 0.5 μm leads to increase in silver clusters size from 800 nm to 2 μm, which consequently causes increase in the overall amplitude of the Raman signal and a decrease in efficiency of SERS spectral resolution. In a sample with an anodic dissolution layer thickness of 0.75 μm, decrease in the size of silver clusters to 800-600 nm was observed. Regardless up to 84% of clusters have a size of 800-600 nm, 2 μm clusters were observed also. We suppose that the cluster size change leads to a decrease or the repeatability of the recorded SERS spectra (for the AP sample with anodic dissolution layer of 0.75 μm value). This may be due to the substrate pores size decrease and their deepening, along with changes in the geometry of the excitation radiation optical rays path on the sample. The experiment demonstrated that the samples had both inhomogeneous structure and significant difference in the degree of the detected signal amplification (up to 41 times), depending on the laser beam position on the surface. Figure 2 illustrates the SERS signal with best resolution, registered with 0.5 μm anodic dissolution value sliver substrate film.

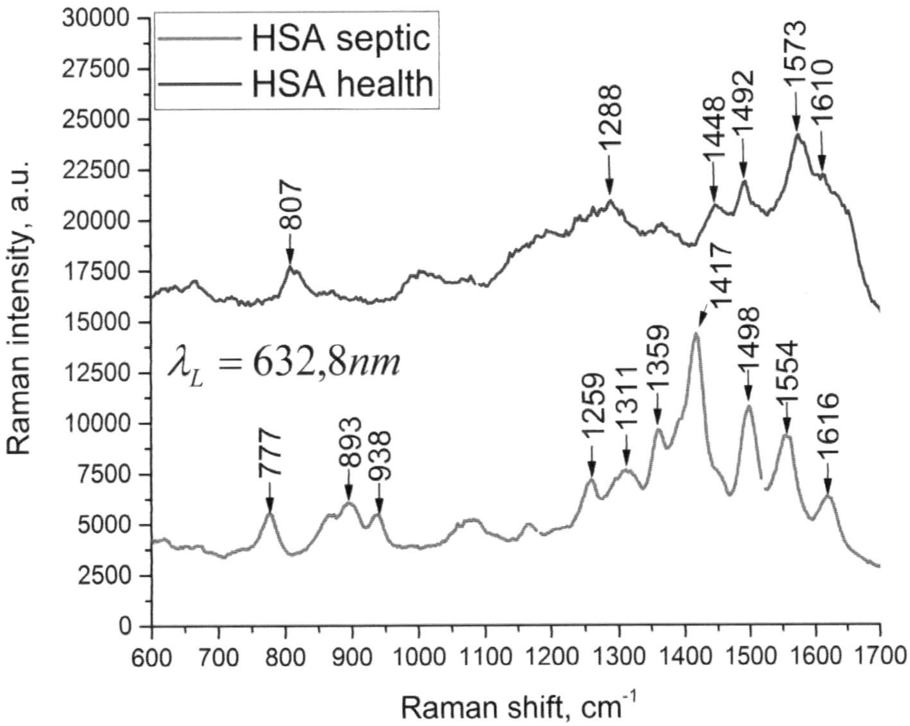

Figure 2. SERS spectra of septic (red line) and health (blue line) HSA on the SERS substrate surface, prepared with anodic dissolution of the silver film with 0,5 μm thick.

Based on the aforementioned study results, the protein structure of healthy patients and the protein of patients with sepsis were identified. The main characteristic frequencies in the frequency range of $\Delta\nu$ = 500-1000 cm^{-1} are the skeletal vibrations of the protein molecule [9], $\Delta\nu$ = 1200-1400 cm^{-1}

denote the regions of the Amide I band [10], Δν = 1600-1700 cm^{-1} denote the regions of the Amide III [10]. Identified vibrational groups are performed in Table 1.

Table 1. Spectral data for septic and health HSA for 0.5 μm anodic dissolution

Protein group	Raman band position of health HSA, cm^{-1}	Raman band position of septic HSA, cm^{-1}	SERS Enhancement factor
Amide III	1288	937, 1259, 1311, 1359	
Amide I	1610	1616	41
Skeletal	807	776, 893	

According to the SERS spectroscopy method, the main frequency bands of the spectrum of the protein molecule were identified and deciphered. According to sepsis pathology, deformation was found to be affected by Amide III group. It structure become more chaotic, the amino acid residue of tryptophan 214 appears on the surface of the protein globule [11], as well as the amino acid residue of tyrosine is packaged inside the molecular structure.

4. Conclusion

The methodology for creating silver films has been developed and optimal experimental parameters for recording the Raman spectra of human serum albumin in sepsis has been performed. It has been shown that the use of the developed substrates with deposited silver films makes it possible to obtain amplified Raman signal. The surface structure with 0.25, 0.5 and 0.75 dissolution AFM results have been presented. Protein molecules structure for normal and sepsic conditions for the main protein vibration groups in the 500-1700 cm^{-1} band has been analyzed.

References

[1] Butler H et al. 2016 *Nature protocols* **11** p. 664-687
[2] Kong K et al. 2015 *Advanced drug delivery reviews* **89** p.121-134.
[3] Freudiger C et al. 2009 *Optics and Photonics News* **12** p. 30.
[4] Liu H et al. 2011 *Scientific Reports* **1**
[5] Kneipp J 2009 *Theoretical Chemistry Accounts* **3-6** p. 319–327.
[6] Vitol E 2012 *Journal of Raman Spectroscopy* **7** p.817–827
[7] Bantz K et al. 2009 *Physical chemistry chemical physics* **24**. p. 11551– 67.
[8] Konstantinova E et al. 2016 *Journal of Physics: Conference Series* **1**. p. 012037.
[9] Kessler J 2015 *The Journal of Physical Chemistry Letters* **16** p. 3314-3319
[10] Lewis E et al. 2014 *Molecules* **12** p.20888-20905.
[11] Kitagawa T and Hirota 2006 *Raman spectroscopy of proteins* **1** p. 2

IOP Conf. Series: Journal of Physics: Conf. Series **945** (2017) 012012 doi:10.1088/1742-6596/945/1/012012

SiPM timing characteristics under conditions of a large background for lidars

A M Antonova, V A Kaplin

National Research Nuclear University MEPhI (Moscow Engineering Physics Institute), Kashirskoe highway 31, Moscow, 115409, Russia

E-mail: Frau.antonova.alexandra@yandex.ru

Abstract. Silicon photomultipliers (SiPM) have found their use in various fields of industry and scientific experiments. This paper considers study of the SiPM possibility to detect low-intensity light pulses (down to single photons) under high-intensity background illumination. This may be useful for the development of laser rangefinders operating under natural light using SiPM as crucial photosensor. Moreover, the presented data describes some physical properties of LIDAR with SiPM under radiation exposure, which always affects its intrinsic noise.

1. Introduction

The determination of the distance to both fixed sites and moving target by means of variable methods is one of the most critical issues of our time. For this purpose, light detection and ranging (LIDAR) are often utilised. In this paper, we consider the possibility of employing silicon photomultipliers (SiPM) in LIDAR. It could be useful for the development of laser rangefinders operating under natural light using SiPM as main photosensor.

SiPMs are very compact devices; it's compatible with complementary metal-oxide semiconductor (CMOS) technology. SiPM production could be low-cost relatively to the convention vacuum photomultipliers. The first prototypes were developed in 1980s-1990s in Russia. The silicon photomultiplier consists of an array of independent cells operating in Geiger mode. The photodetector has the excellent photon-counting capability, high photon detection efficiency, high time, and amplitude resolution; the detector also insensitive to magnetic fields, up to about 7 T. Therefore this type of photodetectors is an advanced instrument used in many applications (e.g. modern physics experiments, accelerators, modern medicine, and outer space), and hence LIDAR with SiPM could benefit a lot for the accuracy of range measurement.

This article outlines the physical properties of the SiPM operations under high-intensity background illumination; herewith it registers one-photon impulses. The problem is relevant for LIDAR, for example that might be related to radiation exposure.

At the beginning of the article a laboratory-scale experimental facility is presented, thereafter, that would be involved. The substance of article contains experiments results and a discussion on the results is presented. The conclusion places capabilities of LIDAR, for which the major component is SiPM, and about the potential applications of this technique.

IOP Conf. Series: Journal of Physics: Conf. Series **945** (2017) 012012 doi:10.1088/1742-6596/945/1/012012

2. The experimental results

2.1. A description of the laboratory-scale experimental facility

The laboratory-scale experimental facility consists of 4 parts: a detecting module (1) with both SiPM and a notch filter; a wideband 1 GHz preamplifier with a gain 100; lens (2); a laser (3) and a reflecting surface (4).

In this case SiPM 1x1mm^2 was used; its peak sensitivity wavelength is 490 nm, the photon detection efficiency (PDE) is 10-12%, the dark count rate (DCR) is from 1 to 3 MHz.

The detecting module registers the laser radiation reflected from a surface. The signal of SiPM with fluctuations can be seen in Figure 2. Background noise flow has roughly equal amplitude. The background noise level had been increased intentionally for research purposes, at least in order to show that SiPM is capable of detecting the reflected laser radiation, even when external noise level is high.

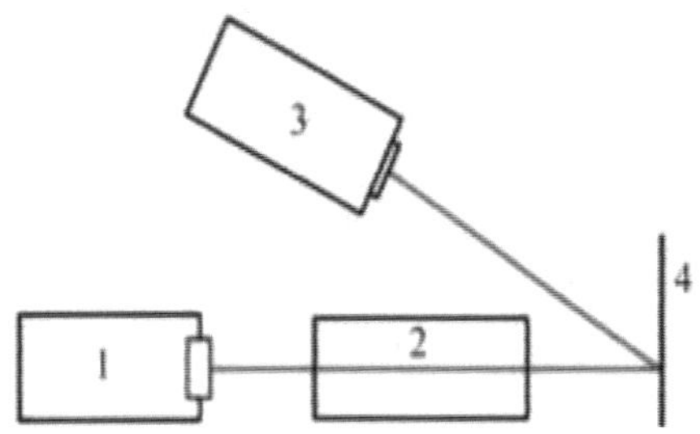

Figure 1. Scheme of the experimental set-up.

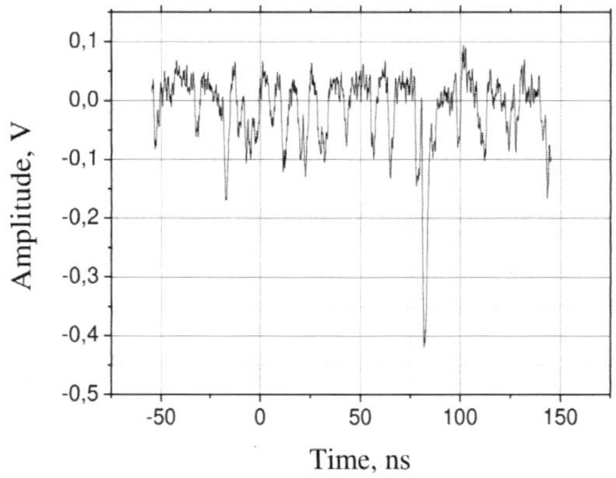

Figure 2. SiPM signal waveform of SiPM with an average background noise pulse rate 100 MHz.

The detector was operated in the linear mode, it can be seen in Figure 3, that shows the verification plot of the single-photon pulse frequency as a function of SiPM current, a linear relationship was observed. Plot is a calibration dependence of the single-photon pulse frequency and here it's needed to guarantee the results of measuring.

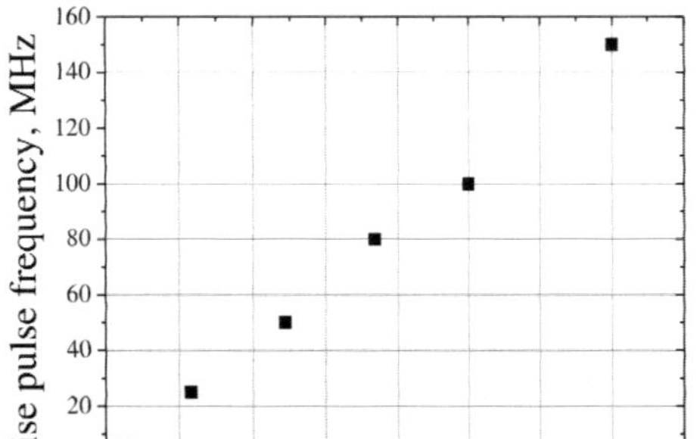

Figure 3. The calibration plot of SiPM current as a function of the noise pulse rate.

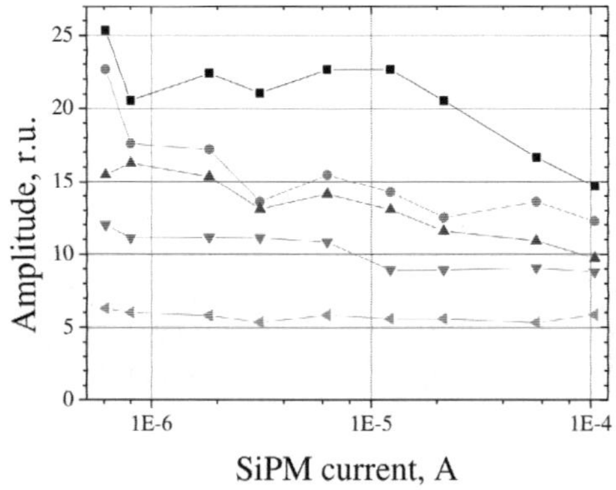

Figure 4. SiPM amplitude vs. current of SiPM.

2.2. An external impact analysis

Further research is needed to analyse the influence of external factors like an external light action, as in this case. The Figure 4 shows the increase number of photon-discharged microcells, when the constant background intensity changes. This may be explicable by the fact that photons are incident on the surface of microcells, when they are activated. If search of the optimum, it needs to take into account, that the amplitude of SiPM signal decreases, when the total number of microcell grows.

Changes of external indirect lighting could affect LIDAR parameters, such as time resolution et al. The Figure 5 provides time resolution as function of average SiPM current, an average number of photon-discharged microcells changes between 6 and 25. As can be seen here, the increasing background illumination leads to decreasing values of time resolution.

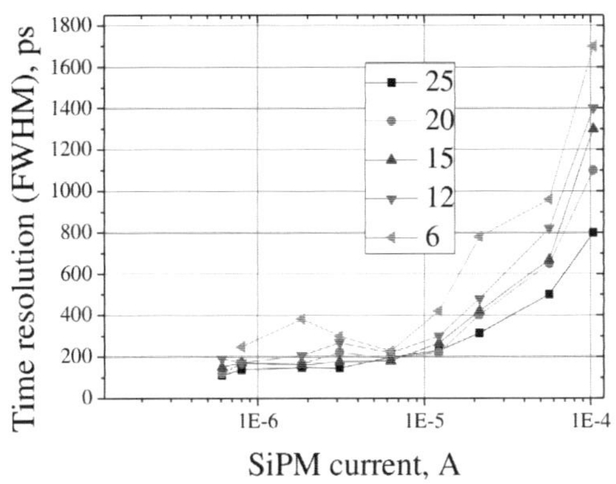

Figure 5. Plot of time resolution vs. current of SiPM for different average number of microcells.

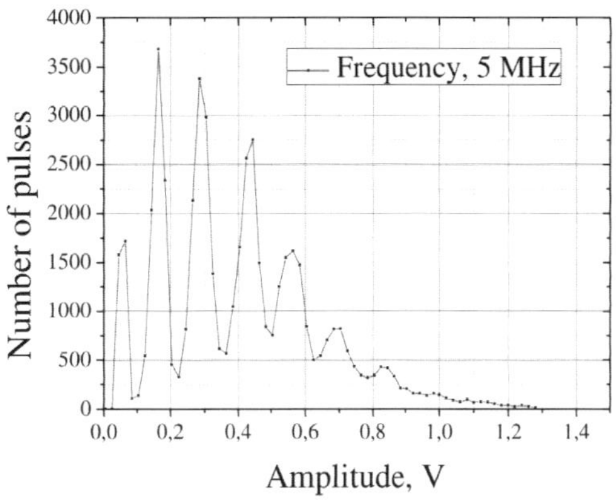

Figure 6. Amplitude spectrum of SiPM.

Oscilloscope measures an amplitude spectrum, as represented in the Figure 6, an average background noise rate is 5 MHz, and the notch filter attenuates the background noise. This is a common spectrum demonstrating the photon-counting capability of SiPM.

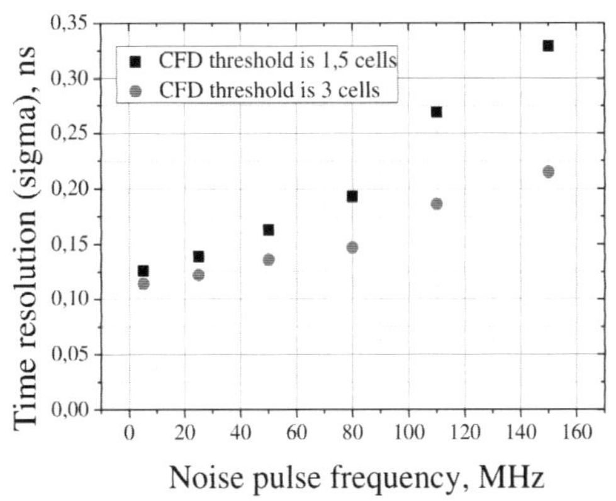

Figure 7. Plot of time resolution vs. a background noise frequency, constant fraction discriminator is used.

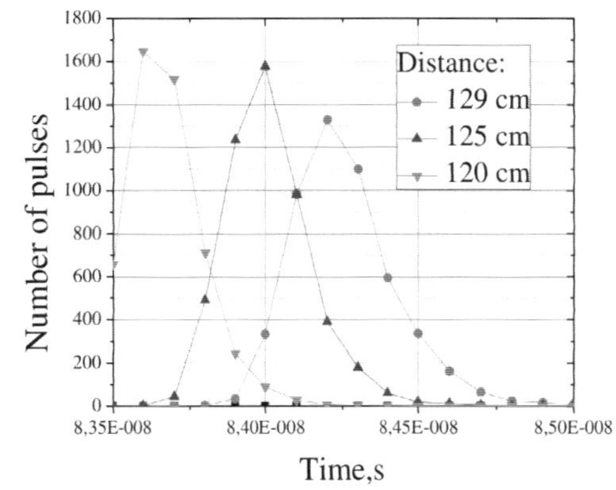

Figure 8. Measuring of distance to rapidly moving objects.

2.3. Time resolution for distance measurements

The accuracy of range measurement to a target can be defined by time resolution. Time resolution starts to deteriorate when the constant fraction discriminator's threshold value is reduced, and it can be seen in Figure 7. Time resolution grows with the increasing of background noise; because, since, statistically, the SiPM signal and the background noise signal could be added to each other.

The Figure 8 indicates the time spectrum, here LIDAR measures the subject distance, changed between 120 cm and 129 cm. Laser pulse duration is 30 ps. In addition, here account should be taken, that 1-5 photons are registered by SiPM in the detecting module. LIDAR has the ability to calculate distances by measuring the time for a signal to return. Therefore, if measurements are made using this method, then the accuracy of LIDAR varies between 5 and 10 cm.

3. Conclusion

The use of SiMP as the substantial photosensor in LIDAR presents clear advantages, such as fast performance and the photon-counting capability along with good time resolution. This device can be used in systems of soft landing on the Moon and in the automobile industry. Furthermore, on this basis 3D – LIDAR could also be created.

Acknowledgements

The authors are grateful to E.F. Maklyaev and Y.A. Melikyan for useful discussions. This work was supported by Competitiveness Program of National Research Nuclear University MEPhI.

References

[1] Agishev R, Comeron A, and Gilerson A 2016 About the potential of liders with different photodetectors under daytime sky radiation (EPJ Web of Conference) **119**

[2] Agishev R, Comeron A, Jordi B, Royo S 2013 Lidar with SiPM: Some capabilities and limitations in real environment (Optics & Laser Technology) **l49** 86–90

[3] Antonova A M and Kaplin V A 2017 Research into the application potential of LIDAR based on SiPM (original Russian text: International Conference for Students "Modern problems of physics and technology") **2** 98-99

[4] Antonova A M 2017 LIDAR based on SiPM (original Russian text: International Conference for Students and Young Scientists "Lomonosov"), p 391-392

IOP Conf. Series: Journal of Physics: Conf. Series **945** (2017) 012013 doi:10.1088/1742-6596/945/1/012013

Characterization of vacuum flash evaporated CdTe thin films for solar cell application

A K Papikyan[1], V A Gevorgyan[1], N R Mangasaryan[1] and P P Gladyshev[2]

[1] Russian-Armenian (Slavonic) University, Armenia,
[2] Dubna State University, Russia

Email: arman.smkl@gmail.com

Abstract. This paper reports the results of CdTe thin films characteristics deposited by vacuum flash evaporation technique. The characteristics of CdTe films deposited at different substrate and evaporator temperatures were analysed using different methods. XRD results have shown that all samples exhibit a predominant sharp peak at 2θ around 23.80 which is assigned to the (111) plane of zinc blend cubic structure. AFM images of CdTe thin films have shown that grain size increases with increasing the substrate temperature, which correlates with XRD results. We have found that the decrease the evaporation temperature from 850^0C to 650^0C leads to the drastically increasing of the average grains sizes from 60 nm to 600 nm. The transmission and reflection spectrum of CdTe thin films are used to evaluate the bandgap Eg. All films have Eg (1.49 eV - 1.50 eV) correspond to the bandgap of a bulk CdTe crystal.

1. Introduction

In last decade cadmium telluride (CdTe) is of great interest as one of the most promising material for low cost and effective solar cells application [1]. CdTe has been considered as one of the most attractive semiconductor absorber layer in thin film solar cells due to its optimal direct bandgap 1.5 eV and high (more than $5{\cdot}10^4$ cm^{-1}) absorption coefficient. A theoretical limit of one-junction CdTe solar cells is more than 30% for AM1.5. [1]

Characteristics of CdTe thin films strongly depend on the applied deposition technique. Various techniques have been reported for deposition of CdTe thin films such as physical vacuum deposition [2], electro deposition [3], closed-space sublimation (CSS) [4] and etc. The highest published in scientific journals efficiency of 16.5% is obtained by using CSS technology [5]. In spite of high value of efficiency, the main disadvantage of CSS technology is the high (650-700^0C) temperature of the process, additionally requires expensive special heat-resistant glass substrates [6]. That is why the further development of lowtemperature methods of CdTe layers deposition is important not only from standpoint of reducing the power consumption, but also for production of solar cells on flexible polymer substrates.

Among the low temperature techniques for manufacturing CdTe solar cells the most suitable are the magnetron spattering [7, 8] and physical vacuum evaporation techniques [9]. The best results were obtained by the method of vacuum thermal evaporation. However, by means of this technique it is difficult to deposit the films of II-VI compounds of strictly stoichiometric composition. From this point of view is more acceptable the flash evaporation technique which is the variation of the thermal vacuum evaporation. Despite this technology advantages there is a limited number of works dedicated to application of this technique for deposition of chalcogenide thin films [10, 11].

As it is described in work [11] the stoichiometric composition of chalcogenide films is achieved because the deposition is performed by discrete evaporation of small portions of alloy to be deposited.

Small particles of crushed chalcogenide alloy whose constituents have different partial vapor pressure fall down into the preheated flash exchanger. Since the particles have different sizes and fall down in a random way at each moment in the flash exchanger there are some amounts of particles which have different temperatures and are at different evaporation stages. Due to the simultaneous presence of such particles in the flash exchanger the composition of deposited film repeats the stoichiometric composition of evaporated material.

In this work, we for the first-time applied vacuum flash evaporation technique to deposit CdTe thin films on glass substrate and investigated the substrate and evaporator temperature's influence on the structural, morphological and optical characteristics of CdTe films which act as an absorbing layer in chalcogenide solar cells.

2. Experimental details

Thin films of CdTe were fabricated by flash evaporation technique on glass substrate with thickness 1 mm and diameter 20 mm, which were cleaned by standard semiconductor technology methods.

The source material was CdTe crystals (99.999%) crushed to a powder with particle size of 50 μm – 150 μm. The vacuum chamber before evaporation was pumped to a base pressure of $1.5 \cdot 10^{-5}$ Torr. Dispensing the CdTe particles from batcher onto the molybdenum flash exchanger was performed due to opening and closing of batcher outlet [11].

Thickness of CdTe films and their deposition time were 1 - 2 μm and 30 - 50 min, respectively.

3. Results and Discussion

3.1. Structural analysis of CdTe thin films

Figure 1 shows the X-ray diffraction patterns of CdTe thin films recorded over 2θ range from 15^0 to 70^0 with the step 0.1^0. All CdTe films deposited at 100^0C, 175^0C and 250^0C substrate temperatures exhibit a predominant sharp peak at 2θ around 23.8^0 which is assigned to the (111) plane of zinc blend cubic structure [ASTM x-ray powder data file, card No. 15-770]. These diffraction patterns show that the (111) orientation increases while decreasing the process temperature, and the results are in good agreement with the same results observed in [12]. The average grain sizes (D) of the film was calculated using the Scherrer's equation [13].

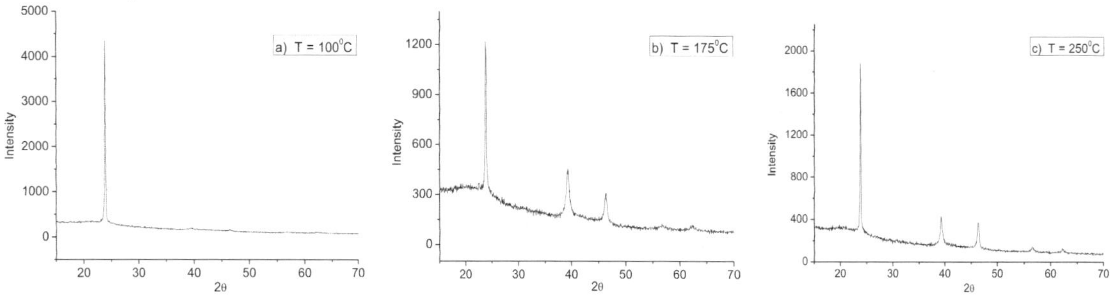

Figure 1. XRD patterns of CdTe films deposited at substrate temperature: a) 100^0C, b) 175^0C, c) 250^0C

The grain size varies with increase the substrate temperature non- monotonically. At first with the temperature increase from 100^0C to 175^0C the grain size decreases from 53.9 nm to 42.1 nm then from 175^0C to 250^0C increases from 42.1 nm to 57.4 nm.

3.2. Optical analysis of CdTe thin films

The results of optical measurements of CdTe thin films deposited at different substrate temperatures are presented in Figure 2 and Figure 3.

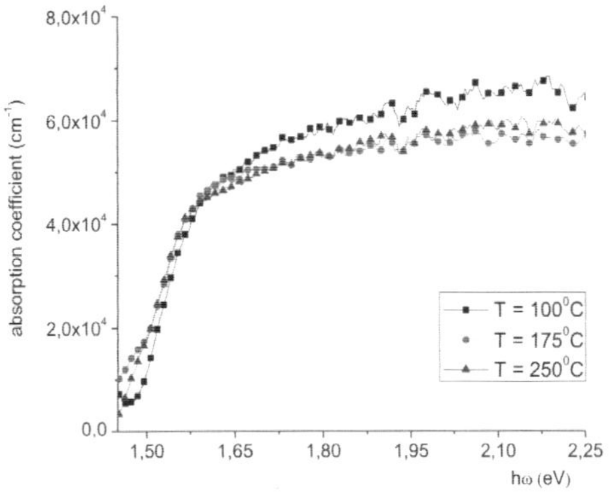

Figure 2. The plot of absorption coefficient (α) vs. photon energy (ħω) at 100^{0}C, 175^{0}C and 250^{0}C substrate temperatures.

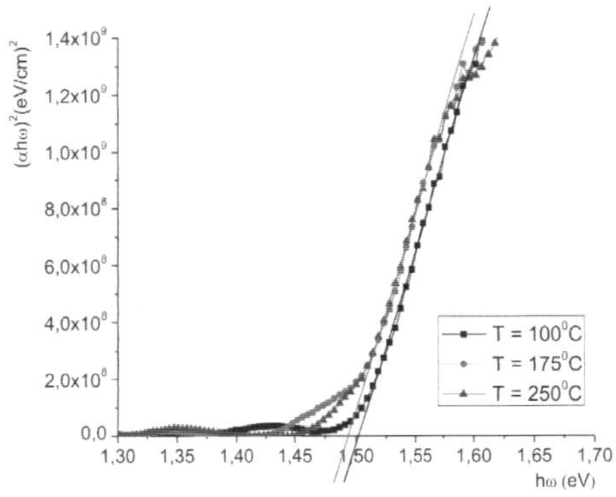

Figure 3. Plots of $(\alpha \cdot \hbar\omega)^2$ vs. photon energy (ħω) at 100^{0}C, 175^{0}C and 250^{0}C substrate temperatures.

The dependence of the absorption coefficient α on the incident photon energy ħω for vertical allowed optical transitions is calculated using Tauc's equation: $\alpha \cdot \hbar\omega = B (\hbar\omega - Eg)^{1/2}$ [14], where Eg is the optical bandgap, B is a constant. Using this equation, the value of CdTe thin films optical bandgap have been calculated and were determined to be 1.50 eV, 1.49 eV and 1.49 eV at substrate temperatures 100^{0}C, 175^{0}C and 250^{0}C, accordingly. The thin film of CdTe deposited at 100^{0}C has the same optical bandgap as the CdTe bulk crystal.

3.3. Surface morphology of CdTe thin films, deposited at different substrate and evaporated temperatures

Figure 4 shows typical 2μm x 2μm AFM images and grain size bar chart of CdTe films deposited on a glass substrate at 250^{0}C and at evaporator temperature 800°C-850°C. The AFM images of films, deposited on substrates having different temperatures showed that substrate temperature increase from 100^{0}C to 250^{0}C leads to only a small increase in grain size (the biggest grain size increased from 120 nm to 160 nm, about 30%). These results are in good agreement with the average grain sizes calculated from XRD results.

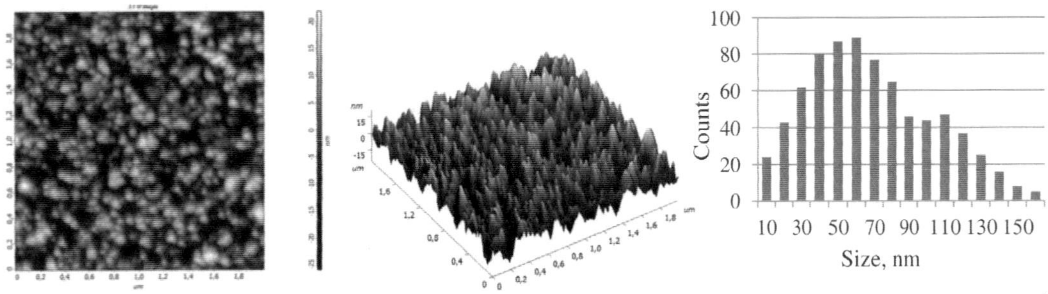

Figure 4. AFM images and the grain size bar chart of CdTe thin films, deposited on substrate at 250^{0}C.

In experiments we have observed, that the grain size depends also on the evaporator temperature. We have found that the grains of huge size (in average 600 nm) are resulting from evaporator temperature

(600°C-650°C) and at substrate temperature 250^{0}C. For comparison, the best results of CdTe thin films deposited by thermal vacuum evaporation technique had the grain sizes of about 100 nm [15]. The AFM images of these films surface morphology and grain size distribution are shown in Figure 5. The XRD results show, that CdTe films deposited at these technological conditions have the best crystalline structure.

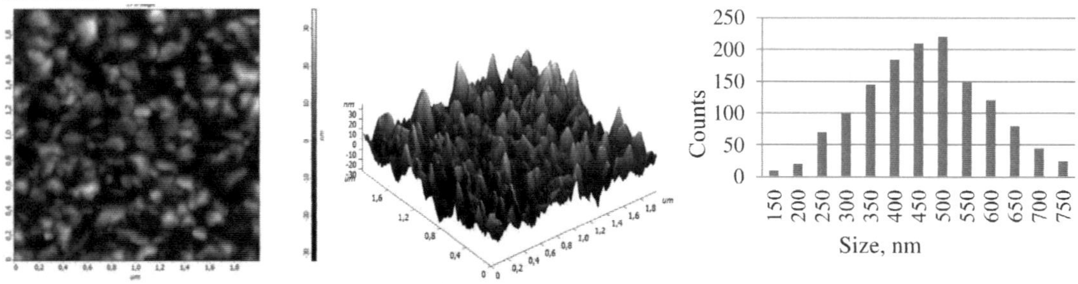

Figure 5. AFM images and the distribution of grain sizes of CdTe thin films, deposited at evaporator temperature 600°C. The substrate temperature was 250°C.

Acknowledgments
The authors are thankful to the Ministry of Education of the Russian Federation for providing financial assistance of this research, as well as to the Competitiveness Program of National Research Nuclear University MEPhI, for organization of conference and publication of its proceedings.

References

[1] Albright S P, Ackerman B, Jordan J 1990 *IEEE Trans. Electron. Devices* **37** (2) p 434.
[2] Khairnar U, Bhavsar D, Vaidya R and Bhavsar G 2003 *Mater. Chem. Phys.* **80** p 421.
[3] Mathew X, Mathews N, Sebastian P and Flores C 2004 *Sol. Energy Mater. Sol. Cells* **81** p 397.
[4] Hernбndez G, Matthew X, Enriquez J, Morales B, Lira M, Toledo J, Juбrez A and Campos J 2004 *J. Mater. Sci.* **39** p 1515.
[5] Wu X, Keane JC, Dheere RG, DeHart C, Albin DS, Duda A, Gessert TA, Asher S, Levi DH and Sheldon P 2001 *Proc. 17th European Photovoltaic Solar Energy Conf.* (Munich), p 995.
[6] Ferekides C S*, Marinskiy D, Viswanathan V, Tetali B, Palekis V, Selvaraj P and Morel D L 2000 *Thin Solid Films* **361**-362 pp 520-6.
[7] Alvin D Compaan, Akhlesh Gupta, Sunghyun Lee, Shanli Wang and Jennifer Drayton 2004 *Solar Energy* **77** Issue 6, pp 815–22.
[8] Mu Y, Zhou X, Yao H, Su S, Lv P, Chen Y, Wang J, Fu W, Song W and Yang H 2015 *J. Alloys Com.* **629** pp 305-9.
[9] Salavei A, Rimmaudo I, Piccinelli F and Romeo A 2013 *Thin Solid Films* **535** pp 257–60.
[10] Gevorkyan V A, Reymers A E, Nersesyan M N and Arzakantsyan M A 2012 *Journal of Physics: Conference Series* **350**, 012027.
[11] Gevorkyan V A, Hakhoyan L A, Mangasaryan N R and Gladyshev P P 2016 *Chalcogenide Letters* **13**, No. 8, pp 331–8.
[12] Birkmire R W and Eser E 1997 *Annual Review of Materials Science* **27** pp 625–53.
[13] Klug H P and Alexander L E 1954 *X-ray diffraction methods for polycrystalline and amophous materials* (New York: John Wiley and Sons).
[14] Tauc J, Grigorov R and Vancu A 1966 *Physica Status Solidi* **15** p 627.
[15] Subhash Chander, Dhaka M S 2015 *Physica E* **73** pp 35–9

Pulsed laser deposition of aluminum nitride nanowires

N R Yunusova, N I Kargin, S M Ryndya, A S Gusev, S V Antonenko and A A Timofeev
National Research Nuclear University «MEPhI», Russia, 115409, Moscow, Kashirskoye sh., 31.

E-mail: naida.yunusova1994@mail.ru

Abstract. The possibility of AlN nanowires deposition on single-crystal silicon substrates by pulsed laser deposition in vacuum is shown in this work. Experimental samples were investigated by scanning electron microscopy and infrared Fourier spectroscopy. It is shown that the possible mechanism for the AlN nanowires formation is the "vapor-liquid-crystal" mechanism.

1. Introduction

A combination of outstanding properties of aluminum nitride, such as a large band gap (6.2 eV), high thermal conductivity (320 W/(m·K) at 300 K), high breakdown field strength (17 kV/mm) and the ability to form a continuous series of AlN - GaN solid solutions, predetermined its wide use in devices of high-power microwave electronics and optoelectronics. The AlN-GaN heterocompositions obtained by molecular-beam epitaxy or gas-phase epitaxy are actively applied for the creation of transistors with 2D electron gas [1], UV light emitting diodes and lasers [2, 3]. In addition, thin AlN films are applicable as dielectric, passivating and heat-conductive layers [4, 5]. However, the most promising structures based on AlN with unique transport, luminescence and adsorption properties today are nanoscale whiskers – a new class of materials for future nanoelectronics and sensors [6, 7]. Therefore, the aim of this work was to study of possibilities of AlN nanowires synthesis on silicon substrates by pulsed laser deposition (PLD).

2. Experiment

Experimental samples of AlN layers were obtained by PLD method from a ceramic AlN target (99.5%) of stoichiometric composition using an excimer KrF laser (laser radiation wavelength 248 nm, pulse duration 20 ns, pulse repetition rate 15 Hz) under high vacuum (3×10^{-6} Torr). Monocrystalline silicon wafers with a diameter of 50 mm and (100) orientation were used as substrates. The samples were grown at substrate temperatures of 600, 700, and 800°C. The substrate heating was carried out by an IR oven. The temperature monitoring was carried out by means of thermocouple and pyrometer. The uniformity of heating was ± 6 °C on all area of a growth surface.

3. Results and discussion

The morphology of a surface of experimental samples was investigated by means of scanning electron microscopy (SEM). Figure 1 shows the SEM images of the planar surface of structures obtained at substrate temperatures (t_{sub}) of 600°C (a), 700°C (b), and 800°C (c). The surface of the sample grown at t_{sub} = 600°C in vacuum (Figure 1, a) is relatively smooth and covered with microdroplets whose diameter fluctuates in the range from 20 to 200 nm, and the surface concentration reaches values of ~10^3 μm^{-2}. The presence of such particles is typical for the films formed by PLD method. Physical mechanisms contributing to the particulates formation are splashing of the molten surface layer caused by surface boiling, expulsion of the liquid layer by shock waves.

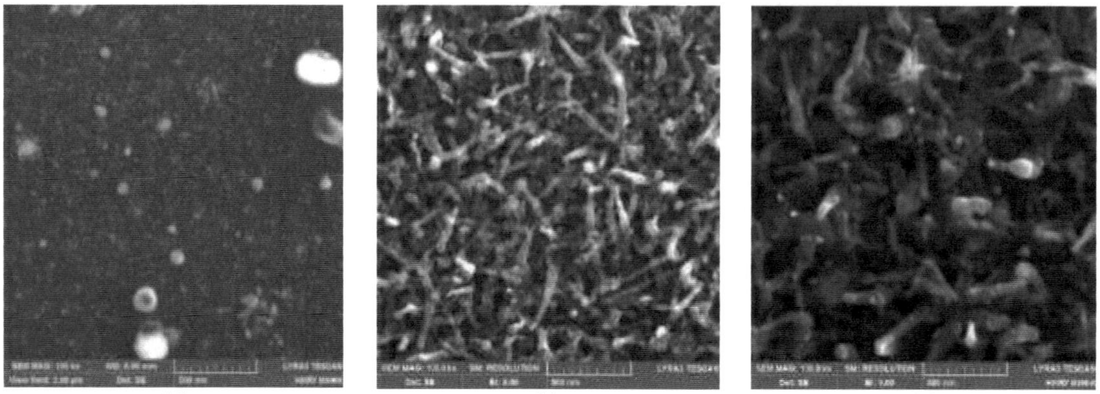

(a) (b) (c)

Figure 1. SEM images of experimental samples deposited at the following substrate temperatures: (a) – 600°C, (b) – 700°C, (c) – 800°C.

In turn, the surface morphology of the sample grown at t_{sub} = 700°C changes drastically (Figure 1, b). The formation of an array of disordered nanowires with a diameter of 35 – 50 nm and a length up to 0.5 μm is observed at this temperature. Also a growth of multiple nanowires from one basis is observed. Nanowires uniformly cover the entire surface of the sample, and there are crystallized droplets in the form of hemispheres at their ends. An increase in t_{sub} up to 800°C leads to an increase of the nanowires diameter and to decrease in their surface density (Figure 1,c).

In figures 2,a and 2,b the SEM images of the cross section of the samples deposited under the substrate temperatures of 600°C and 700°C are shown. It can be seen from these figures that the film thickness for the sample without AlN nanowires is about 100 nm. The height of the nanowires formed on the surface of the sample at 700°C reaches a value of 400 nm (Figure 2,b), and the crystallized droplets at the ends of the nanowires sometimes have a slightly pronounced facet (Figure, 2c).

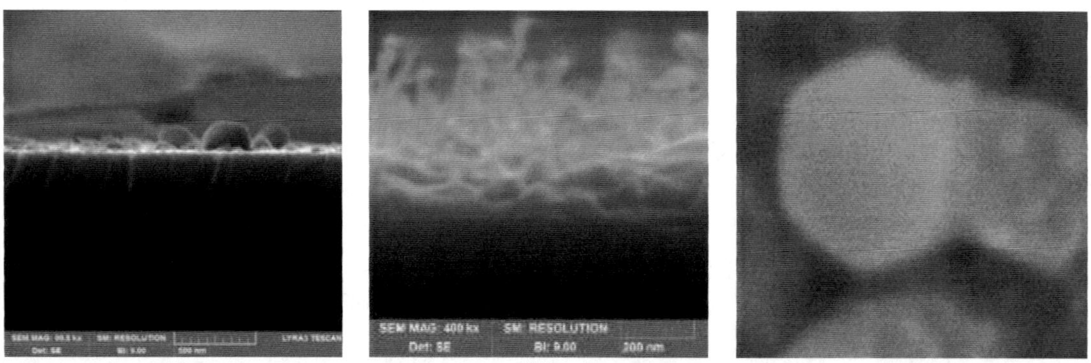

(a) (b) (c)

Figure 2. SEM images of the cross-section of samples deposited at substrate temperatures of: (a) – 600 °C, (b) – 700 °C; (c) – SEM image of individual AlN nanowire.

An array of oriented AlN nanowires of similar dimensions (with a diameter of 50-60 nm and a length of about 250 nm) were obtained by molecular beam epitaxy [8] on a single-crystal silicon substrate. However, the mechanism of formation of such nanowires in [8] is not disclosed.

In order to evaluate the composition and structural features of the formed layers, spectra of their transmission in the IR region were obtained (Figure 3, a). Measurements of the transmission spectra were carried out by 8400S (Shimadzu) FTIR spectrometer using standard technique with a resolution of 2 cm^{-1}. It can be seen from Figure 3a that the spectra of experimental samples deposited at substrate temperatures of 700 and 800°C contain one broad absorption band in the range of 400 – 1000 cm^{-1}, which can be associated with the vibrational modes of the AlN lattice [9]. Moreover, as the t_{sub} parameter increases, the intensity of this band increases, and its half-width decreases from 200 cm^{-1} (at t_{sub} = 700°C) to 130 cm^{-1} (at t_{sub} = 800°C). The decrease in the half-width of the absorption band with increasing temperature is associated with an improvement in the crystallinity of the layers being formed. In turn, the sample obtained at 700°C demonstrates strong absorption in the entire measured spectral range, which is probably related to the accumulation of metal (Al) in its volume.

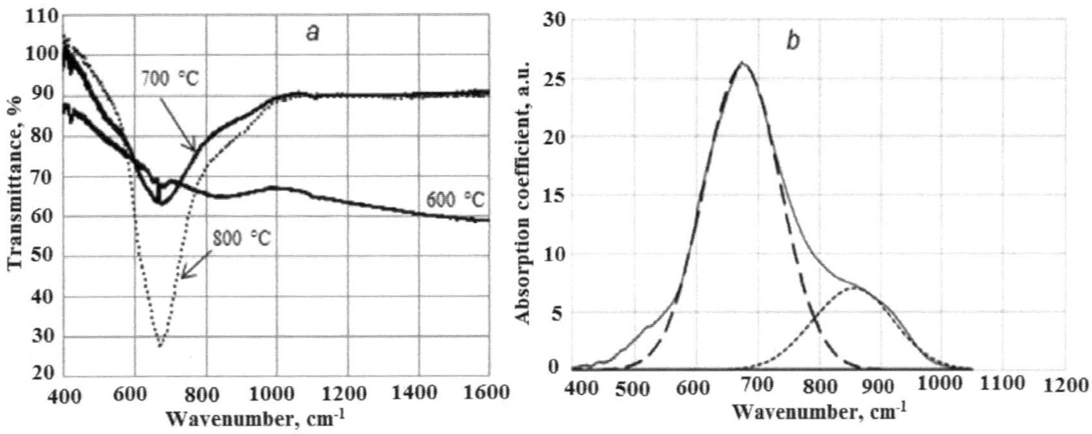

Figure 3. IR-transmission spectra of experimental samples (a). The result of decomposition of the absorption spectrum for the sample deposited at 800°C (b).

For the sample deposited at 800°C, the transmission spectrum T (%) was recalculated in the spectral distribution of an absorption coefficient α (cm^{-1}) followed by the decomposition of the integral spectra into two Gaussian components (Figure 3, b), the parameters of which are presented in Table 1. According to literature data, the transverse optical phonon mode of AlN (TO) lies in the range of 660 – 670 cm^{-1}, the longitudinal optical modes of AlN (LO) are in the range of 880 – 915 cm^{-1} [10, 11]. Thus, the first peak with ω_{max} = 674 cm^{-1} can be attributed to the TO vibration mode of Al-N bonds. However, the second peak with ω_{max} = 855 cm^{-1} is located in the range between TO and LO modes. This peak can be associated with the appearance of surface modes of optical oscillations (SO) at crystallite sizes less than the wavelength of incident light [10]. The observed size effect confirms formation of AlN nanowires.

Table 1. Parameters of absorption spectrum

Peak №	Peak position, cm^{-1}	Peak half-width, cm^{-1}	Intensity, rel. units	Description
1	674	146	26	TO – transverse optical phonon mode of AlN
2	855	156	7	SO – surface optical mode of AlN NW

The possible mechanism for the AlN nanowires formation is the "vapor-liquid-crystal" (VLS) mechanism. Excess Al accumulates on the growth surface during the formation of the AlN layer under

IOP Conf. Series: Journal of Physics: Conf. Series **945** (2017) 012014 doi:10.1088/1742-6596/945/1/012014

vacuum. This excess Al is collected to the nanodroplets, which react with the substrate dissolving silicon in its volume. Then Al-Si melt is acting as a catalyst for the growth of nanowires [12]. The presence of crystallized droplets at the ends of whiskers and the absence of nanowire at t_{sub} = 600°C (that is lower than temperature of Al melting) confirm this assumption.

4. Conclusion

The samples containing arrays of the disordered AlN nanowires on single-crystal silicon substrates were obtained at t_{sub} = 700 and 800°C by PLD method. The experimental samples were investigated by scanning electron microscopy and infrared Fourier spectroscopy. It is shown that the possible mechanism for the AlN nanowires formation is the "vapor-liquid-crystal" (VLS) mechanism.

This work was carried out using the equipment of MEPhI Shared-Use Equipment Center «Heterostructural microwave electronics and wide band gap semiconductor physics».

References

[1] Fedorov Yu 2011 Wide-gap heterostructure (Al,Ga,In)N and device for millimeter waves *Electronics: Science, Technology, Business* № 2 pp 92 – 107 (in Russian).

[2] Feng Z Ch 2008 III-Nitride Devices and NanoEngineering *Imperial College Press.*

[3] Aokia Y, Kuwabara M and Yamashita Y 2015 A 350-nm-band GaN/AlGaN multiple-quantum-well laser diode on bulk GaN *Appl. Phys. Lett.* **107** 151103.

[4] Hirano M, Kato K, Isobe T and Hirano T 1993 Sintering and characterization of fully dense aluminium nitride ceramics *Journal of Materials Science* **28** 4725–30.

[5] Puryzhinskiy S Z, Pebalk D V, Kalashnikov A Y 2015 Microstructural studies of the surface of the aluminum nitride boards from various manufacturers *In the World of Scientific Discoveries* № 8 216–231 (in Russian).

[6] Li Y, Xiang J and Qian F 2006 Dopant-Free GaN/AlN/AlGaN Radial Nanowire Heterostructures as High Electron Mobility Transistors *Nano Letters* **6** №. 7 1468 –73.

[7] Weinstein I A, Chaikin D V, Spiridonov D M, Vokhmintsev A S and Afonin Yu D 2014 Photoluminescence of oxygen complexes in aluminum nitride nanowhiskers *VESTNIK VSGUTU* №6 (in Russian).

[8] Zhao S, Connie A T, Dastjerdi M H T, Kong X H, Wang Q, Djavid M, Sadaf S, Liu X D, Shih I, Guo H and Mi Z 2015 Aluminum nitride nanowire light emitting diodes: Breaking the fundamental bottleneck of deep ultraviolet light sources, *Scientific Reports* **5** № 8332 1–5

[9] Jagannadham K, Sharma A K and Wei Q 1998 Structural characteristics of AlN films deposited by pulsed laser deposition and reactive magnetron sputtering: A comparative study *J. Vac. Sci. Technol. A* **16** № 5 2804–15.

[10] Li H D, Zou G T and Wang H 1998 Synthesis and Infrared Study of Nanosized Aluminum Nitride Powders Prepared by Direct Current Arc Plasma *Journal of Physical Chemistry* **102** 8692 –95.

[11] Jouan P Y; Tricoteaux A and Horny N 2006 Elaboration of nitride thin films by reactive sputtering *Revista Escola de Minas* **59** №. 2 pp 225 – 232

[12] Dubrovskii V G, Cirlin G E and Ustinov V M 2009 Semiconductor nanowhiskers: Synthesis, properties, and applications, *Semiconductors* **43** № 12 (in Russian).

IOP Conf. Series: Journal of Physics: Conf. Series **945** (2017) 012015 doi:10.1088/1742-6596/945/1/012015

The influence of cooling on the magnetic properties of the second generation HTS tapes

I V Anischenko, S V Pokrovskii, I A Rudnev

National Research Nuclear University MEPhI (Moscow Engineering Physics Institute), Kashirskoe sh. 31, 115409 Moscow, Russia

mephizic@gmail.com

Abstract. In this work it is presented a computational model of a magnetic system based on stacks of high-temperature second generation superconducting tapes (HTS) GdBa2Cu3O7-x. The model is designed to calculate magnetic systems based on HTSC tapes taking into account the thermal processes occurring in the layered structure of the tape. The dependences of the trapped magnetic flux and the cooling power on the temperature for different magnetization durations were obtained, and the dynamics of thermal processes taking in the tape during magnetization was analyzed.

1. Introduction

Currently, HTS tapes are used extensively in the manufacture of trapped flux magnets, HTS bearings, levitation suspensions and other levitation devices. The magnetic properties and levitation characteristics of the tapes stacks depend not only on critical characteristics of the used materials but its geometry, conditions of magnetization and cooling. The local heat jumps in the superconducting material can contribute the transition between superconducting state and normal state, complete degradation of the trapped magnetic flux, and, as a consequence, the failure of the working device. Therefore, it is necessary to predict the behavior of high-temperature superconducting tapes in an external magnetic field, taking into account the thermal processes occurring within their layered structure. The present work is devoted to the development of a numerical model that makes it possible to calculate the parameters of a magnetic system based on HTS tapes taking into account heating processes in the system.

2. Computational model

The measurements were performed on a superconducting tapes produced by SuperOx (Russia) with the width of 12mm, thickness of the superconducting layer GdBa2Cu3O7-x is 1 micron and the critical current - 300 A (criterion 1 μV / cm, 77 K, own field). The simulated system is a layered structure of the superconducting tape containing a substrate, a superconducting layer, and also a stabilizing layer of copper. By the others layers we neglect because of their small thickness and small influence on the thermal processes. Cooling within the model is carried out by a cupper bulk, which is a cryocooler analog. The outer boundary of the cryocooler, which is in direct contact with the HTS tape substrate, is set as a cooling source with a determined constant temperature. In the study, we calculated the power, corresponding to the real cooling system power, which is necessary to hold the set temperature in the system. The sample is in a vacuum. The tape is placed on a copper bulk that will be considered as a cooling source. An external uniform magnetic field, 2T of magnitude, linearly increases and then similarly linearly remove out of the system.The geometry of the system is shown in the Figure 1.

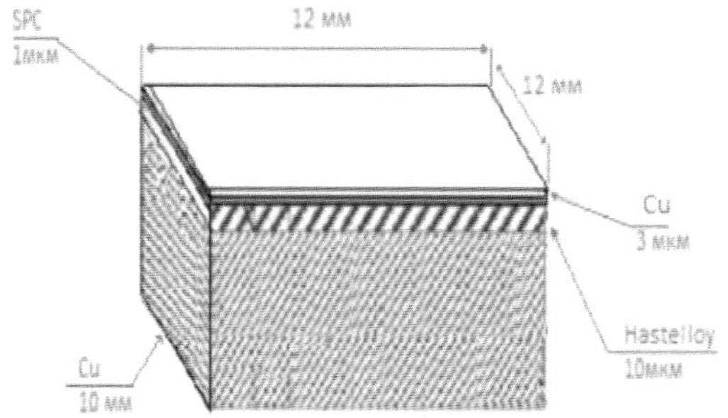

Figure 1. The geometry of the magnetic levitation system.

To describe the superconducting magnetic system we used Faraday and Ampere laws equations. The resistance nonlinear dependence on the current is given by the power law (1) and the current-voltage characteristic can be written by formula (2):

$$\rho = \frac{E_c}{J_c} \cdot \left(\frac{|J|}{J_c}\right)^{n-1} \quad (1)$$

$$\boldsymbol{E} = \rho \cdot \boldsymbol{J} \quad (2)$$

,where E_c is the electric field which equals 1 μV/cm, that corresponds to the critical current. The n-value is assumed to be 21, which is typical for the samples of HTS tapes. The critical current density J_c is determined on the basis of the transport measurements of the used HTS tapes.

The model was developed on the base of the finite element method, which was implemented using the Partial Differential Equations (PDE) form of the Comsol Multiphysics software module. In Comsol Multiphysics the general PDE standard equation view is (3):

$$e_a \frac{\partial^2 \boldsymbol{u}}{\partial t^2} + d_a \frac{\partial \boldsymbol{u}}{\partial t} + \nabla \times \Gamma = F \quad (3)$$

Here the dependent variable, magnetic field $\boldsymbol{u} = \boldsymbol{H}(x,y,z)$, is chosen in the entire space. The parameters of equation (3) are chosen in such a way that equation becomes Faraday's law. Specifications of the parameters of the equations (3) see [1].

Also, in this model it is taken into account the dependence of the critical current on the magnetic field perpendicular to the wide surface of the tapes [2] and experimental dependence of the critical current on the temperature, that described by formula (4):

$$J_{c0} = \alpha\left(1 - \left(\frac{T}{T_c}\right)^2\right)^{1.5} \quad (4)$$

, where α - fitting parameter.

The description of the heat transfer physics is based on the expression:

$$\rho C_p \frac{\partial T}{\partial t} + \nabla \cdot (-k\nabla T) = Q \quad (5)$$

,where C_p-heat capacity, ρ-density, k- coefficient of thermal conductivity of the used materials. In the right-hand side of equation (5), all heat sources Q, including thermoelectric, located. In our case, in the absence of convective heat transfer, the main source of heating is the superconducting tape. Local heat generation in the system is calculated as:

$$Q = E \cdot J \quad (6)$$

,where E is the electric field inside the superconductor. The current distribution $\boldsymbol{J}$ of the superconductor is determined by the conditions of magnetization and is calculated as the solution of the Maxwell equations, realized in the PDE form. In addition, in modeling, the dependences of the heat capacity and the thermal conductivity on the temperature for the superconducting layer were taken into account [3]. For other tape layers, these coefficients were laid in the form of averaged values, characteristic for the selected temperature range. This assumption is correct due to a small temperature change of these layers when the tape is heated

3. Results and discussion

The simulation of tape magnetization in an external uniform magnetic field for different magnetization velocities was carried out. The magnetization times were 0.5s, 2s, 10s, 30s, 60s the cooling temperatures – 40K, 50K, 60K, 65K, 80K.

As expected, at lower temperatures the HTS sample traps larger magnetic field (Figure 2a). The maximum value of the trapped magnetic flux H_{max}= $6.4*10^5$A/m (0.8T) at a height of 1 mm from the surface of the tape for the magnetization duration 0.5s at 40K. However, more intensive magnetization regime, leading to more intense heating, requires higher cooling power (Figure 2, b).

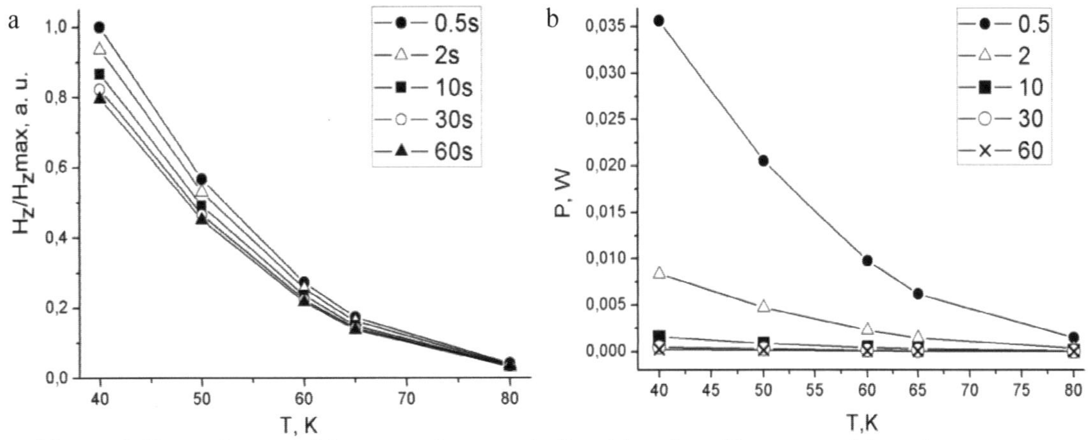

Figure 2. Dependences of the trapped magnetic flux (a) and cooling power (b) on the temperature.

Temperature maintenance for the short-pulse magnetization requires powers several orders higher, than for the long-pulse magnetization, while the trapped flux value increases only about 20%. This indicates the existence of an optimal magnetization regime for trapping the flux under the given magnetization and cooling conditions.

In addition, increasing of the magnetization velocity also leads to an increase in the trapped magnetic flux. Similar results for magnetization by single pulses were obtained experimentally [4]. Such magnetic behavior can be explained by the thermal processes occurring in the tape during the magnetization. At slow magnetization, heat flux has time to spread throughout the sample. Temperature increasing leads to a decrease in the critical current of the sample, and, consequently, to trapped flux decreasing. At short-pulsed magnetization, on the contrary, the thermal spot, hat appears on the edges of the tape, does not have time to spread into the interior of the sample and does not affect its magnetic characteristics.

Figure 3a shows the temperature gradient dependences on the magnetization time for the 40K and 80K temperatures (applied field 0.5T). It can be seen that when magnetizing with short pulses There are local hot regions with a temperature jump several times greater than in the case of slow magnetization. At the same time, at a lower temperature, a larger gradient appears.

Using the calculation model, it can be seen that the magnetic field begins to enter the HTS tape from the edges, where the maximum heat release occurs at the initial times (Figure 3, b). At the initial magnetization stages, the magnetic field Hz and the gradient temperature field permeate the sample practically at the same time. Then the resulting thermal spot begins to propagate through the sample due to the own thermal conductivity of the superconductor and the stabilizing layers. The main heat outflow is through the substrate [5].

Copper coating contributes to the dissipation of the sharpest maximums of heat, thus, copper performs a stabilizing function. Ultimately, thermal processes becomes steady and new temperature bursts do not arise. The heat release along the edges of the tape continues until the magnetization process is completed.

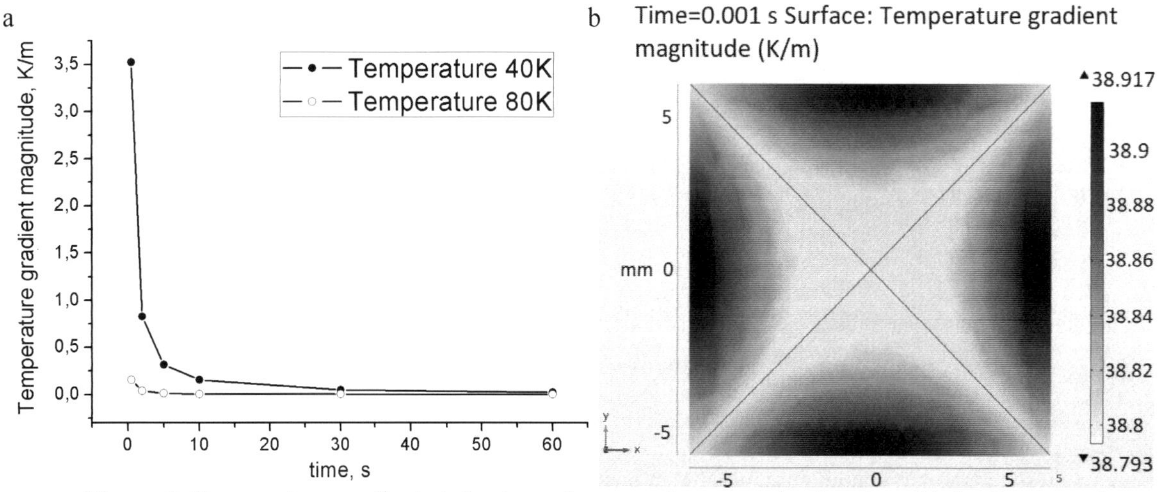

Figure 3. Temperature gradient a) the dependence on the magnetization time, b) the "heat spot" appearance

4. Conclusions

Based on the use of the finite element method the physical model of the magnetic levitation system element based on the second generation HTS tape was developed. The model uses the features of the tape's layered structure. For the superconducting layer, the dependences of the heat release, cooling power, and trapped magnetic flux on the temperature for different magnetization rates were obtained.

Obtained results indicate the existence of optimal parameters of temperature and magnetization velocity to achieve the maximum performance of devices based on HTS tapes stacks.

Acknowledgments

The study was carried out at the expense of a grant from the Russian Science Foundation (Project No. 17-19-01527).

References

[1] Rodriguez-Zermeno V.M., Sørensen M.P., Pedersen N.F., Mijatovica N. and Abrahamsen A.B 2009 *Fast 2D Simulation of Superconductors: a Multiscale Approach* Proceedings of the COMSOL conference. Vol. User presentations and proceedings CD, pp 1-6

[2] Podlivaev A.I., Rudnev I.A. and Shabanova N.P. 2014 Magnetic-Field Dependence of the Local Critical Current Density in the Second-Generation HTSC Tapes *Bulletin of the Lebedev Physics Institute* **41** №. 12 pp 351–4.

[3] Zhang M., Matsuda K. and Coombs T.A. 2012 New application of temperature-dependent modelling of high temperature superconductors: Quench propagation and pulse magnetization *Journal of Applied Physics* **112** 043912

[4] Moses A.J. and Basak A. 1995 *Nonlinear electromagnetic systems: proceedings of the*

International ISEM Symposium on Nonlinear Electromagnetic Systems Cardiff, Wales, UK, 17-20 September, ed. A.J. Moses and A. Basak pp 116-9

[5] Page A.G., Patel A., Baskys A., Hopkins S.C, Kalitka V., Molodyk A. and Glowacki1 B.A. 2014 The effect of stabilizer on the trapped field of stacks of superconducting tape magnetized by a pulsed field *Superconductor Science and Technology* **28** №. 8

IOP Conf. Series: Journal of Physics: Conf. Series **945** (2017) 012016 doi:10.1088/1742-6596/945/1/012016

The possibilities of dimensional electron-beam processing as applied to selective sintering of oxide ceramics in the fore-vacuum pressure range

I Yu Bakeev[1], E S Dvilis[2], A S Klimov[1], E M Oks[1,3] and A A Zenin[1]

[1]Department of Physics, Tomsk state university of control systems and radioelectronics, Tomsk, 40 Lenina Street, 634050, Russia
[2]Innovation Centre for Nanomaterials and Nanotechnologies, Tomsk polytechnic university, Tomsk, 30 Lenina Street, 634034, Russia
[3]Laboratory of plasma sources, Institute of High Current Electronics, Siberian Branch, Russian Academy of Sciences (IHCE SB RAS), Tomsk, 2/3 Akademichesky Avenue, 634055, Russia

E-mail: bakeeviyu@mail.ru

Abstract. The paper presents the results of studies devoted to an effect of deviation and scanning parameters on the value of power density of a focused electron beam generated by a plasma electron source in the forevacuum pressure range. It is shown that the change in the beam power density does not exceed 20% with an angle of beam deviation within 20 degrees in the investigated pressure range of 6-8 pascals in air. The results of selective sintering of powder system of zirconia-alumina by an electron beam generated by a forevacuum plasma electron source are demonstrated.

1. Introduction

Ensuring of effective precision electron-beam processing of products is largely connected with the achievement of high power density of the electron beam in a local area [1]. One of the known examples of the precision application of an electron beam is the EBM technology, which is used to create products with an arbitrary shape from titanium alloys [2]. However, as shown in [3], when irradiation of metal powders is used in this technology at working pressures of 10^{-1}-10^{-2} Pa, there is a problem of spread of the powder particles caused by the process of their charging by the electrons. The solution of this problem consists in preheating the volume of powder to the sintering temperature of metal particles. The usage of high-temperature ceramic powders for the synthesis of three-dimentional products is complicated by the need to attract some special measures to neutralize the accumulated negative charge [4]. In this case, it is advisable to use a forevacuum plasma source [5], which generates an electron beam at the pressures from one to hundreds of pascals. During working in this pressure range, almost complete neutralization of the accumulated negative charge is realized by a flow of ions from the plasma formed by ionization of residual gas by the beam electrons and ignition of a glow discharge between the irradiated surface and walls of a vacuum chamber [6, 7].

Effective focusing of an electron beam in the region of high pressures in the forevacuum range is primarily complicated by the processes of scattering of the electrons on molecules of residual gas.

Despite this, we were able to obtain the electron beam with a diameter less than 0.6 mm and provide the power density of 10^4-10^5 W/cm^2, which was sufficient to "drill" holes in high-temperature ceramics [8]. However, in order to achieve a high quality of electron-beam synthesis, it is necessary to preserve specific parameters of the electron beam effect when a deflection of the beam is occur. Thus, the purpose of the investigations was to demonstrate the possibility of direct electron-beam dimensional processing by the forevacuum plasma electron source and investigate the possibility of selective synthesis of a ceramic powder of the zirconia-alumina system by this source.

2. Scheme and technique of experiment

The scheme of the experimental setup, the technique of electron-beam sintering and the measurement of beam sizes are shown in Figure 1. The electron beam *1* is generated by the forevacuum plasma electron source *2*, whose operation principle is based on emission of electrons from plasma of a glow discharge with a hollow cathode. The parameters of the electron beam, which were used in the experiments, are the following: the accelerated voltage Ua is12 – 16 kV, the beam current Ib is up to 15 mA; the pressure in the vacuum chamber is controlled by feeding working gas (air) and is within P = 6-8 Pa. In more detail, the design and features of functioning of this electron source are presented in [9]. The magnetic lens *3* performs focusing of the electron beam. Using the deflection system *4* consisting of two pairs of magnetic coils, the beam deflection from the original direction and its scanning are performed. The current signal of the deflection system is controlled by an electronic system for formation and deflection of an electron beam designed and manufactured at NPT TETA [10] for electron-beam welding setups.

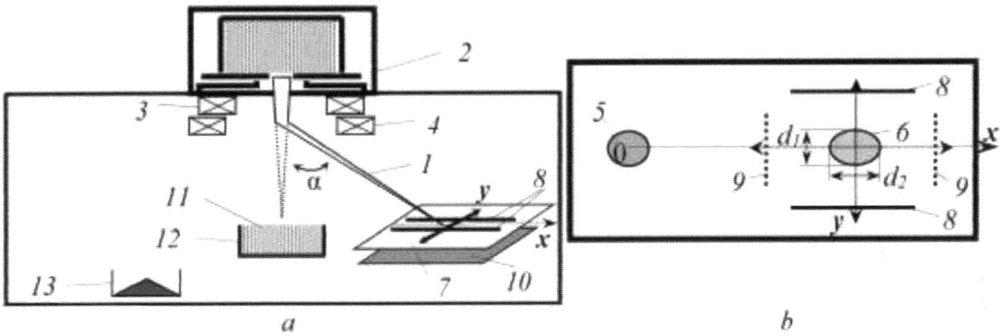

Figure 1. Scheme of the experiment and illustration of the technique of electron beam size measuring (a – scheme of beam size measuring and powder sintering; b – beam measurement plane (top view): 1 – electron beam; 2 – forevacuum plasma electron source; 3 – focused magnetic lens; 4 – deflection system; 5 – position and cross-section form of electron beam without deflection; 6 – position and cross-section form of electron beam when it is deflected at angle α; 7 –horizontal plate of the beam size measurement system; 8 – pair of gaps for measuring minimum size of beam d_1; 9 – pair of gaps for measuring maximum size of beam d_2; 10 – collector for beam current measuring; 11 – sintered ceramic powder; 12 – graphite crucible; 13 – Faraday cylinder; x, y – directions of beam deflection.

The investigation of cross-sectional size change of the electron beam during its deviation by an angle α (Figure 1 a) is carried out by the "deflection" method [11]. Due to changing the form of electron beam cross-section from round (α = 0, *5* in Figure 1 b) to oval (α > 0, *6* in Figure 1 b), the beam power density is determined by equation: $q = (4 \cdot U_a \cdot I_b)/(\pi \cdot d_1 \cdot d_2)$, where d_1, d_2 – minimum and maximum cross sizes of the beam respectively, which are measured in the mutually perpendicular directions (Figure 1 b).

The measurements of the cross sizes of the beam is carried out by the following way: the electron beam is focused on the plate *7* which is perpendicular to the propagation direction of beam, then the electron beam is deflected by an angle α along the ***x*** axis (Figure 1 a) from the initial position. The electron beam consistently intersects one of two pairs of the narrow measuring gaps *8*, *9* with the width of 0.2 mm located on a special displacement device. One pair of gaps is located along the $\boldsymbol{x}$ axis, the other is perpendicular to it along the $\boldsymbol{y}$ axis. The beam electrons that have passed through the gaps are received by the collector *10*, a current signal from which is fed to an oscilloscope. When the beam intersects the gaps *8* along $\boldsymbol{x}$, the beam size d_1 is determined. Accordingly, the size d_2 is determined during crossing the pair of gaps *9* along the $\boldsymbol{y}$ axis. The values of d_1 and d_2 are estimated as $d_{1,2} = L{\cdot}\tau/T$, where L – the distance between gaps, T – time interval between amplitude values of the signal on the oscillogram, τ – peak width on the oscillogram at half-height.

The electron-beam sintering of the powder *11* (zirconia-alumina system with a weight rate 6:1 the average particle size is 1-10 μm) located in a cavity of the graphite crucible *12* in the free-filling state is performed by the beam scanned the surface of the powder in the form of a raster with sides 10x10 mm with the frequency of 60 Hz. To avoid appearance of cracks in the volume of the sintered sample, controlled warming and cooling after processing are implemented with a rate of change of power density 2 W/cm^2·min. The dwell time at the fixed beam parameters during sintering is 30 minutes.

The morphological features and elemental composition of a surface of the sintered samples were studied by scanning electron microscopy (SEM) and energy dispersive spectrometry (EDS) at the JSM-7500FA (JEOL, Japan).

3. Experimental results

As it is shown in [8], the increase in pressure unambiguously leads to the beam broadening. It is obvious that this influence is expressed to a greater extent with the larger scattering cross section of gas molecules. In particular, it is shown in Figure 2 that when the air pressure in the working chamber is increased from 6 to 8 Pa, the beam power density decreases almost fivefold due to the increase in beam geometric sizes. However, despite the scattering processes that negatively affect on focusing of the electron beam, when the beam deflection is performed within 20 degrees, the reduction in the power density does not exceed 20% from the initial value. Such small decrease in the power density of narrow-focused beam indicates the possibility of efficient dimensional precision processing of high-temperature dielectrics.

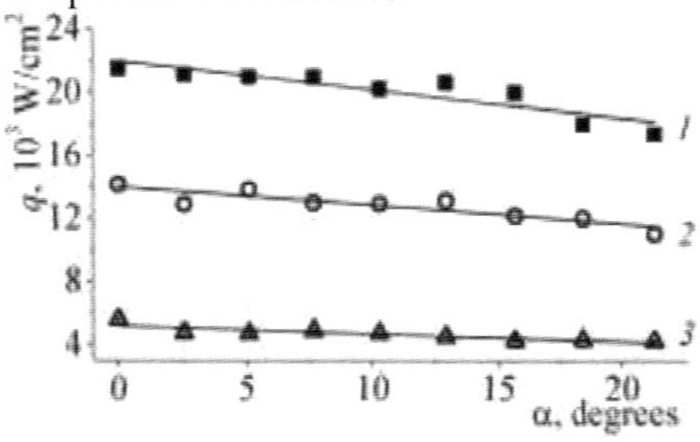

Figure 2. Dependences of the electron beam power density q on the deflection angle α for different gas pressures P (U_a = 16 kV, I_b = 5 mA, working gas – air): 1 – P = 6 Pa; 2 – P = 7 Pa; 3 – P = 8 Pa.

As a demonstration of the effectiveness of using the narrow-focused electron beam generated by the forevacuum plasma source, the ceramic powder of the zirconia-alumina system with a weight ratio of 6: 1 was selectively sintered. The SEM images of surface of the sintered samples with different integrated power densities and EDS data are presented in Figure 3. Analysis of the obtained results and their comparison with the known ratio of the components of the initial powder mixture (6: 1), allow us to conclude that significant alumina segregation occurs on the pore surfaces or the zirconia particles after the action ofthe electron-beam sintering. This process is quantitatively expressed by the conditional value of segregation coefficient k of the alumina, which is defined as the ratio of the excess founded by the EDS method to the zirconia content of the initial mixture. Figure 3.1 shows that the part of alumina is sintered in a solid phase with insufficient intensity of the impact, and the effects of grain-boundary slip during sintering of zirconia particles are not completely realized. In the case of excessive intense melting of the alumina, there is a restriction of the sintering zirconia particles together (Figure 3.3). In this case, the zirconia particles are separately distributed in a liquid matrix of the alumina, but their amount is not sufficient to form a continuous ceramic layer on the surface of the sample.

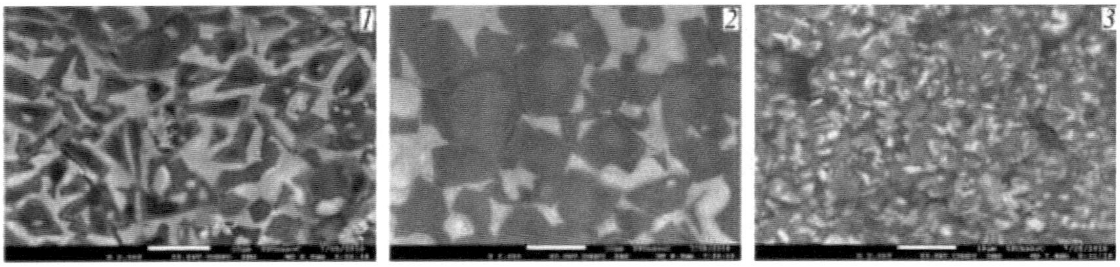

Figure 3. SEM images of surfaces of the sintered ceramics samples, obtained in backscattering electrons: light gray areas – ZrO_2; gray areas – Al_2O_3; dark gray fields – deepening of surface relief and pores: *1* – I = 80 W/cm^2, Ua = 12 kV, C Al2O3 = 42,99 mas.%, C ZrO2 = 57,01 mas.%, k = 0,33; *2* – I = 125 W/cm^2, Ua = 12 kV, C Al2O3 = 53,66 mas.%, C ZrO2 = 46,34 mas.%, k = 0,46; *3* – I = 170 W/cm^2, Ua = 12 kV, C Al2O3 = 54,42 mas.%, C ZrO2 = 45,58 mas.%, k = 0,47

In the case of the optimal mode for the electron-beam sintering (Figure 3.2), the surface has dense interphase boundaries and a complex grain structure. In this structure, in addition to individual equiaxed grains of aluminum oxide and ensembles of extended grains of zirconia, two-phase grains with a core of zirconium dioxide encased in an alumina shell are observed.

4. Conclusions

Thus, the results of the investigations indicate the possibility of efficient dimensional processing of high-temperature dielectrics by the electron beam generated by the plasma source of electrons in the forevacuum pressure range. In addition, the possibility of selective electron-beam sintering of a zirconia-alumina powder is demonstrated.

Acknowledgments

The researches of the influence of deviation processes on beam parameters are supported by the RFBR grant No. 16-38-60032 m_dc. The investigations devoted to electron-beam sintering of ceramic powders are supported by the RFBR grant No. 15-08-00871 a. The research of the influence of gas pressure on electron beam scattering is supported by the Ministry of Education and Science of the Russian Federation within the grant 3.9605.2017/8.9.

References

[1] Rykalin N N, Uglov A A, Zuev I V and Kokora A N 1985 *Laser and electron-beam processing of materials* (Moscow, Mashinostroenie) [in Russian]
[2] Heinl P *et al* 2008 *Acta biomaterialia* **4(5)** 1536
[3] Sigl M, Lutzmann S, and Zäh M F 2006 *Solid Freeform Fabrication Symposium Proceedings* (Austin, TX) p 397
[4] Filachev A M and Fouks B I 1996 *Prikladnaya fizika* **3** 39
[5] Burdovitsin V A, Klimov A S, Medovnik A V, Oks E M and Yushkov Y G 2014 *Forevacuum Plasma Electron Sources* (Tomsk, Izd. Tomsk. univer.) [in Russian]
[6] Burdovitsin V A, Klimov A S, Medovnik A V and Oks E M 2010 *Plasma Sources Science and Technology* **19** 055003
[7] Zenin A A and Klimov A S 2013 *Doklady Tomskogo Gosudarstvennogo Universiteta Sistem Upravleniya I Radioelektroniki* **1(27)** 10
[8] Zenin A A, Bakeev I Y, Burachevskii Y A, Klimov A S and Oks E M 2016 *Technical Physics Letters* **42(7)** 712

[9] Burdovitsin V A, Zhirkov I S, Oks E M, Osipov I V and Fedorov M V 2005 *Instruments and Experimental Techniques* **48(6)** 761
[10] TETA [Electronic resource]: official site of LTD «SPC TETa». *URL: http://tetacom.ru/*
[11] Kaur A, Ribton C, Balachandaran W 2015 *Journal of materials processing technology* **221** 225

IOP Conf. Series: Journal of Physics: Conf. Series **945** (2017) 012017 doi:10.1088/1742-6596/945/1/012017

Modeling of the optimal transmission spectrum of an interference contrast filter using simulated annealing

E Brick[1], S Ermakov[2], S Kucheryavyy[2] and A Furminskiy[2]

[1]Photooptic Inc., 249039, Kaluga Region, Obninsk, Aksenova str, 6B, Russia
[2]National Research Nuclear University MEPhI,115409, Russia, Moscow, Kashirskoe shosse, 31

E-mail: furminskiy@gmail.com

Abstract. The problem of searching for the optimal transmission spectrum of an interference contrast filter for an arbitrary display is considered. It relates to problems of global optimization and is solved numerically using simulated annealing.

1. Introduction.

The problem of readability of information from all kinds of displays in conditions of high external illumination is significant in a modern technological society.

One of the solutions to this problem is the use of interference contrast filters of various types [1], in particular, multi-layer absorbing contrast filters. Such filters are widely used in aeronautical engineering, which makes the task of developing such filters relevant. Currently, the application of contrast neutral filters prevails, i.e. filters with a constant level of absorption over the entire visible range. Such filters are often not optimal by a level of contrast ratio. The transmission spectrum of the contrast filter is an important parameter for the problem under consideration. We are succeeded in adjusting it so that it would be best suited for the selected display, and gives the best contrast ratio. There are models for calculating [2] the contrast ratio of the display with the contrast filter.

The problem of finding the optimal transmission spectrum of a contrast filter is a problem of global optimization. The optimized function has several extrema and is complex for finding the global maximum analytically, therefore the problem is solved numerically using simulated annealing [3][4].

2. Investigated function.

The problem we are solving is based on the model for calculating the contrast ratio proposed by AVAGO engineers [2]. According to the proposed model, the contrast ratio can be calculated as:

$$Cr = 1 + \frac{\sum\left[W(\lambda)Y(\lambda)T(\lambda)\right]}{\sum\left[Sun(\lambda)Y(\lambda)\left(Rf(\lambda) + Rd(\lambda)T^2(\lambda)\right)\right]}, \quad (1)$$

where the sum is over the entire visible region ($\lambda \in$ [400 nm, 720 nm.]) in steps of λ equal to 5 nm.

In this equation:

$W(\lambda)$ — the emission flux density of the display,
$Y(\lambda)$ — the eye perception curve of the average observer,
$T(\lambda)$ — the transmission spectrum of the contrast filter,
$Rf(\lambda)$ — the spectrum of diffuse reflection from the surface of the contrast filter,
$Rd(\lambda)$ — the spectrum of diffuse reflection from the display surface,

Sun (λ) — the external emission flux density.

Contrast-neutral filters currently used can be calculated from the simplified model as described in [2]. For them, equation (1) can be reduced to the form:

$$Cr = 1 + \frac{\pi \cdot L_{disp} \cdot T}{E \cdot (R_{disp} \cdot T + R_{filter})}, \qquad (2)$$

where:
L_{disp} — the own brightness of the display segment,
R_{disp} — diffuse reflection from the display surface without a filter,
R_{filter} — diffuse reflection from the surface of the filter,
E — external illumination,
T_f — the transmission of a gray filter in which $dT / d\lambda = 0$.

The maximum of the contrast ratio in this case will be determined by the filter transmission value equal to $T_f = \left(\frac{R_{filter}}{R_{disp}}\right)^{\frac{1}{2}}$.

A contrast-neutral filter with such a transmission level will be optimal for displays which emission spectrum is not constant. However, for displays with a constant emission spectrum, it makes sense to consider the general case (1).

3. Using simulated annealing.

Simulated annealing, as the name suggests, has a physical analogy, when the lowest potential energy in a metal is achieved with slow cooling.

The main advantage of this method is that with correctly chosen parameters and a sufficient number of iterations, it allows finding the global minimum of the function being investigated. The sensitivity to the form of the function and the choice of the initial point of this method is inconsiderable.

We introduce following notation:
D — the dimension of the space on which investigated function is given,
T_0 — the initial value of the mathematical analog of temperature in the method,
x — a point in a multidimensional space on which the value of the investigated function depends,
x' — the point at which the transition is made to the algorithm iteration,
$E = f(x)$ — the value of the function at the point, the mathematical analog of the potential energy,
ΔE — the change in energy upon transition from point x to point x'.

The specific scheme of the simulated annealing is given by the following parameters:

- The choice of the law of variation of the analogue of temperature $T(k)$, where k is the step number;
- The choice of the generating family of distributions $Q(x, T)$;
- The choice of the probability of acceptance of $h(\Delta E, T)$.

We realized the Boltzmann annealing [3]. It is characterized by the following parameters:
Temperature change

$$T(k) = \frac{T_0}{\ln(1+k)}, k > 0 \qquad (3)$$

A family of distribution with a distribution density $g(x';x,T) \in Q(x, T)$

$$g(x', x, T) = (2\pi T)^{-\frac{D}{2}} \exp\left(-\frac{|x' - x|^2}{2T}\right) \qquad (4)$$

Acceptance probability function

IOP Conf. Series: Journal of Physics: Conf. Series **945** (2017) 012017 doi:10.1088/1742-6596/945/1/012017

$$h(\Delta E, T) = \exp\left(-\frac{\Delta E}{T}\right) \quad (5)$$

It was shown in [4] that for sufficiently high initial temperature and total number of steps, choosing this family of distribution guarantees localization of a global extremum.

An algorithm for modeling optimal transmission spectra of interference contrast filters for arbitrarily selected displays whose emission spectrum is known has the following form [3]:

1) The initial point $x=x_0$ is randomly chosen. The current value of the energy E is set to the value $f(x_0)$.

2) the k-th iteration of the main cycle consists of the following steps:

- Compare the energy of the system E in the state x with the currently found minimum. If $E = f(x)$ is smaller, then change the value of the global minimum.
- Generate a new point $x' = G(x, T(k))$ where $G(x, T(k)) \in Q(x, T)$ is the chosen distribution function.
- Calculate the value of the function in this point $E' = f(x')$
- Generate a random number α from the interval [0; 1]
- If $\alpha < h(E'-E, T(k))$, then set $x = x'$, $E = E'$ and proceed to the next iteration. Otherwise, return to generating a new point.

We have written a program that is based on this algorithm with the help of which calculations were made.

We have calculated the transmission spectra for various possible emission spectra of displays with the assumption that they have a diffuse reflection value $Rd = 10\%$, and the contrast filters have $Rf = 0.04\%$. Figure 1 represent two model spectra for which the calculation has been made.

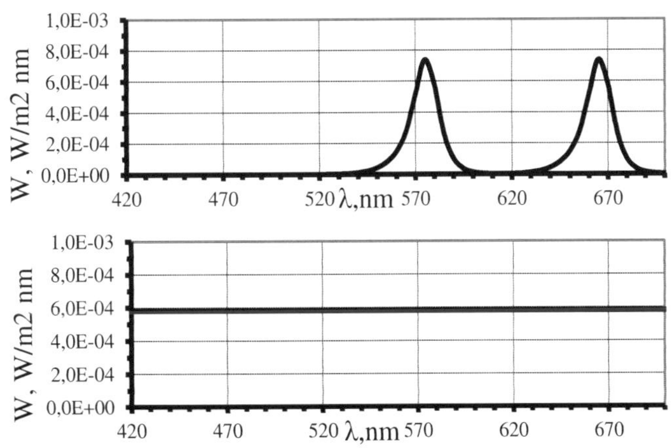

Figure 1. Model emission spectra of displays.

For them, the following values of the best contrast ratio $Cr = 5.43$, $Cr = 11.51$ have been calculated in accordance with (1) with the following optimal transmission spectra.

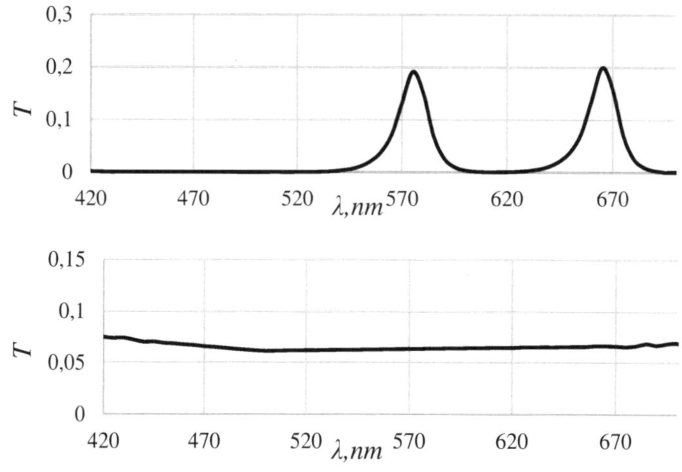

Figure 2. Optimal transmission spectra of hypothetical contrast filters in the appropriate order

4. Conclusions

We have measured the contrast ratio using the available contrast filters to compare the experimental data with the calculated values obtained with usage of the proposed model. The accuracy of coincidence of the results with the calculations is ~ 5%.

Using the implemented algorithm, an optimal transmission spectrum was obtained for the display of seven segmented indicators with the emission spectrum presented below.

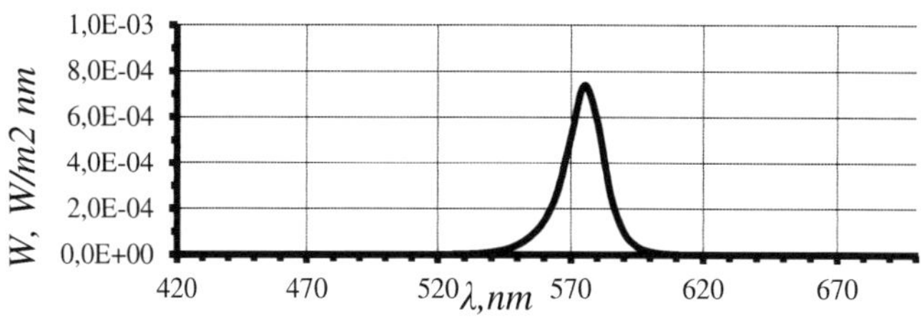

Figure 3 The measured emission spectrum, for which the optimal transmission spectrum of the contrast filter was modeled.

And the following parameters:

The overall brightness level of the display — $L = 1620\frac{cd}{m^2}$

Diffuse reflection from the display surface — $Rd = 10\%$

The following results were obtained for him:

The optimal transmission for contrast neutral coating and the corresponding contrast ratio obtained by (1) and (2) at Rf=0.04% is T=20% and Cr=2.16.

The optimal transmission spectrum of the contrast filter for this display is as follows:

IOP Conf. Series: Journal of Physics: Conf. Series **945** (2017) 012017 doi:10.1088/1742-6596/945/1/012017

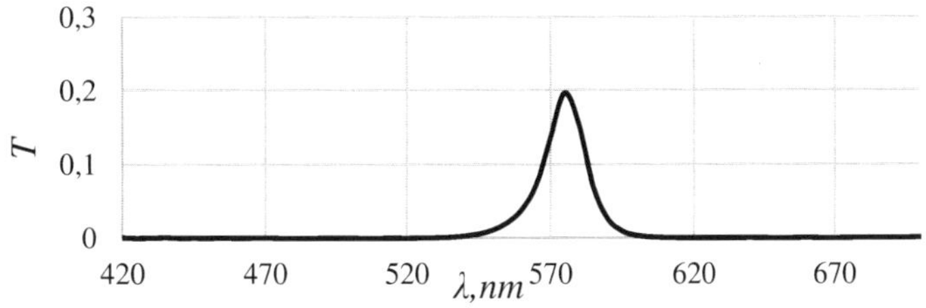

Figure 4 The optimum transmission spectrum of the contrast filter, for the display with the parameters described above.

For this configuration, the contrast ratio calculated by (1) is Cr=5.31.

As can be seen, the solution of considered problem allows, for displays with a constant spectrum of radiation, to find the transmission spectrum of the contrast filter, which, when installed, gives about a two and a half times more contrast ratio than with standard contrast neutral filters.

References

[1] Born M Wolf E 1970 *Principles of optics. Fourth edition* – Pergamon Press p 808
[2] AVAGO Contrast Enhancement Techniques for LED Displays Application Note 1015.2016
[3] Lopatin A S 2005 *Simulated annealing* – Saint Petersburg University. Saint Petersburg p 148
[4] Metropolis N Rosenbluth A W Rosenbluth M N Teller A H and Teller E 1953 Equation of State Calculations by Fast Computer Machines *J. Chemical Physics* **21** p 1087-1092

IOP Conf. Series: Journal of Physics: Conf. Series **945** (2017) 012018 doi:10.1088/1742-6596/945/1/012018

Two-photon coherent spectroscopy of ultracold Li atoms

A A Bobrov[1], E V Vilshanskaya[1,3], B B Zelener[1,4], S A Saakyan[1] and V A Sautenkov[1,2]

[1] Joint Institute for High Temperatures, Russian Academy of Sciences, Izhorskaya Street 13, Bd. 2, Moscow 125412, Russia
[2] P. N. Lebedev Physical Institute, Russian Academy of Sciences, Leninskii Prospect 53, Moscow 119991, Russia
[3] National Research University - Moscow Power Engineering Institute, Krasnokazarmennaya 17, Moscow 111250, Russia
[4] National Research Nuclear University - Moscow Engineering Physics Institute, Kashirskoye Chaussee 31, Moscow 115409, Russia

E-mail: abobrov@inbox.ru

Abstract. Our work is devoted to theoretical study of the two-photon coherent spectroscopy of ^{7}Li atoms continuously cooled in a magneto-optical trap (MOT) on the 2S–2P transition. The ultracold atoms are transferred to highly excited Rydberg states in a two-step coherent excitation process by red and UV lasers. The red laser is detuned by -600 MHz from 2S–2P transition frequency and UV laser frequency detuning is scanned in the vicinity of +600 MHz from 2P-nS(D) transition where n∼40-100 is principal quantum number. The fluorescence signal on the 2P–2S cooling transition makes it possible to obtain a two-photon absorption spectrum. Atom-field interaction is considered in the simple three-level approximation involving a density matrix formalism. It is shown that the effect of the MOT beams on the shape of the two-photon absorption line can be taken into account by an appropriate change in the 2S–nS(D) coherence decay rate.

1. Introduction

Investigation of the properties of systems consisting of highly excited Rydberg atoms is important for various applications, such as, for example, quantum computations [1, 2].

It is convenient to use two-photon or three-photon laser excitation to transfer atoms to Rydberg states. The standard method for diagnosing Rydberg states uses ionization of excited atoms by an electric field [3]. For a simple and reliable diagnosics of Rydberg states in the two-photon excitation process, we proposed a new method based on recording the resonance fluorescence of atoms in the MOT [4, 5, 6]. The advantages of this method include simplicity and high sensitivity. High sensitivity is due to the low rate of escape of atoms from the MOT and the low atoms capture rate in the MOT. Small rates determine small nonlinearities that ensure saturation of the signal.

In this paper we propose a simple theoretical model that describes well the shape of the spectral line of the two-photon absorption, and also makes it possible to estimate the densities of atoms in the ground and Rydberg states.

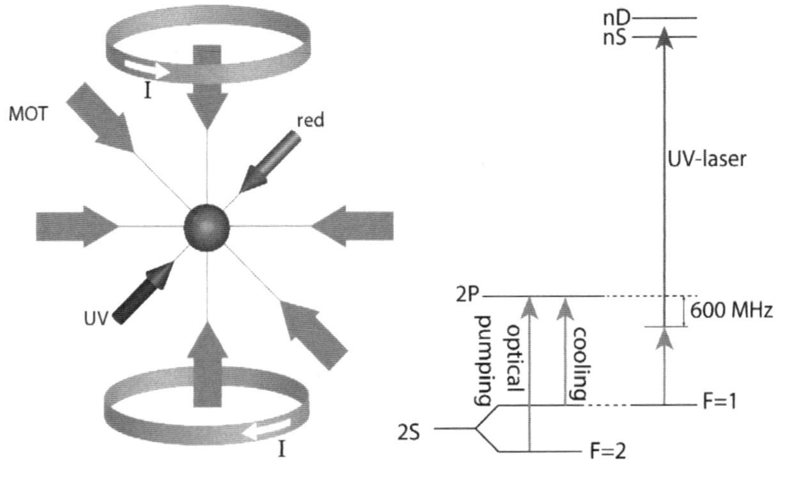

Figure 1. Schematic diagram of the experiment and involved ^{7}Li levels.

2. Experiment

A schematic diagram of the experiment on two-photon excitation of ultracold ^{7}Li atoms and the involved atomic levels are shown in figure 1. The Li atoms are continuously cooled in the MOT at the 2S(F=1)–2P transition, optical pumping is performed at the 2S(F=2)–2P transition. The temperature of the atoms is in the range $T \sim 300 - 500$ μK (depending on the power of the MOT beams). The experimental cooling technique is described in more detail in [7]. The cloud of ultracold atoms is irradiated by counter propagating beams of red and UV lasers. The red laser is detuned from the transition frequency 2S(F=1)–2P to $\delta_1 \sim 600$ MHz. The detuning δ_2 of the UV laser is scanned in the vicinity of +600 MHz from the 2P–nS(D) transition, where n is the principal quantum number of the corresponding Rydberg level of Li. By tuning the frequency of the UV laser, n can be set within wide limits n$\sim$40–100.

At close absolute values of the detunings of the red and UV lasers, a two-photon coherent absorption process occurs, leading to the formation of Rydberg Li atoms with definite values of the principal n and orbital l quantum numbers. This, in turn, leads to an effective decrease in the density of atoms in the ground state and an experimentally observed drop in fluorescence at the 2P–2S transition (see figure 2).

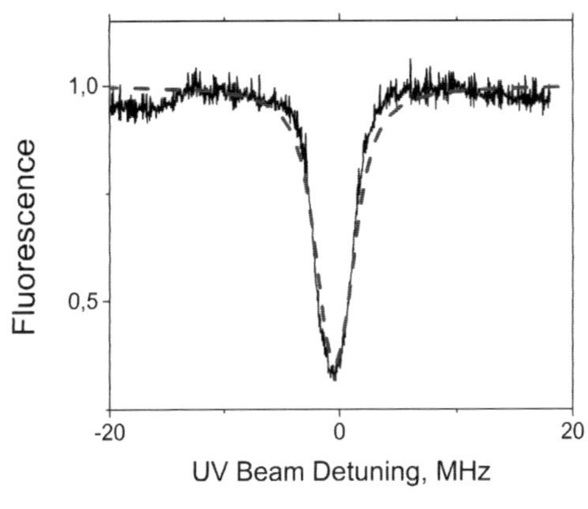

Figure 2. The fluorescence signal of 2P–2S transition as a function of the detuning of the UV laser from the 2P–58S transition (shifted so that the center of the line is approximately at zero). The solid line is the experimental result, the dashed line is the solution of equations (1). The line FWHM is 3.5 MHz, the total intensity of the MOL beams is 8.4 mW/cm^2.

In the next section, we propose a simple model that describes well the experimental results obtained.

3. Theoretical model and comparison with the experiment

For a theoretical description of the experimental results, we will use the density matrix formalism.

Since the relative population of the intermediate level 2P is low [4] and large absolute values of detunings of the red and the UV lasers are used, the Rydberg states are populated during a purely coherent two-photon absorption. This allows us to consider a three-level system consisting of 2S(F=1), 2P and nS(D) states, on which only the fields of the red and the UV lasers operate. In this case, the direct action of the MOT beams on the kinetics of the population of highly excited levels can be neglected.

We denote the levels 2S(F=1), 2P, nS(D) as 1, 2 and 3 respectively. We denote the transitions 2S(F=1)–2P and 2P–nS(D) as 1 and 2 and introduce for them the Rabi frequencies $\Omega_i = d_i E_i/\hbar$, where d_i and E_i are the dipole matrix elements of the corresponding transitions and amplitudes of the radiation fields of red and UV lasers.

We will consider an open system, in this case the diagonal terms of the density matrix ρ correspond to the relative populations of the levels. Following [8], we write the equations for the elements of the density matrix in the three-level approximation in the following form:

$$\begin{aligned}
\dot{\rho}_{11} &= 0.5i\Omega_1(\rho_{21} - \rho_{12}) + 2\gamma_2\rho_{22} + \gamma_{trap}(1 - \rho_{11}) \\
\dot{\rho}_{22} &= 0.5i\Omega_1(\rho_{12} - \rho_{21}) - 2\gamma_2\rho_{22} + 0.5i\Omega_2(\rho_{32} - \rho_{23}) \\
\dot{\rho}_{33} &= 0.5i\Omega_2(\rho_{23} - \rho_{32}) - 2\gamma_3\rho_{33} - \gamma_R\rho_{33} \\
\dot{\rho}_{12} &= 0.5i\Omega_1(\rho_{22} - \rho_{11}) - \rho_{12}(\gamma_2 + i(\delta_1 - k_1 v)) - 0.5i\Omega_2\rho_{13} \\
\dot{\rho}_{23} &= 0.5i\Omega_2(\rho_{33} - \rho_{22}) - \rho_{23}(\gamma_3 + \gamma_2 + i(\delta_2 + k_2 v)) + 0.5i\Omega_1\rho_{13} \\
\dot{\rho}_{13} &= 0.5i\Omega_1\rho_{23} - 0.5i(\delta_1 + \delta_2 - (k_1 - k_2)v)\rho_{13} - 0.5i\Omega_2\rho_{12} - \rho_{13}\gamma_{13} \\
\rho_{ij} &= \rho^*_{ji}
\end{aligned} \tag{1}$$

Relaxation terms containing the corresponding rates are added to the equations: $2\gamma_2$ is the rate of radiative decay of the 2P level, γ_{trap} is the rate of escape of the 2S atoms from the trap, $2\gamma_3$ is the rate of radiative decay of the Rydberg state, γ_R is the rate of escape of Rydberg atoms from the trap, and γ_{13} is the rate of decay of the coherence between the ground and Rydberg states.

Unlike experiments with heavier elements, in the analysis of experiments with Li we cannot neglect the Doppler broadening. The Doppler shifts in (1) for red and UV lasers are denoted by $k_1 v$ and $-k_2 v$ respectively.

Equations (1) are written for the case when the radiative decay of a Rydberg atom removes it from the system. This approximation is made, since in the described experiments there is no radiation recorded that indicates that during the radiation decay the atom could return to the ground state and be again captured in the MOT. The characteristic rates of radiative relaxation processes can be estimated from the extrapolation of data of [9]: for the 40S state $2\gamma_3 \approx 2 \times 10^4$ s^{-1}, for the 40D $2\gamma_3 \approx 3 \times 10^4$ s^{-1} state.

One of the features of the described experiments is that due to relatively low Li atomic mass, the capture rate of atoms in the trap is much less than the escape rate of Rydberg atoms (which are not held by the MOT forces), i.e. $\gamma_{trap} \ll \gamma_R$. The capture rate of atoms can be estimated from the rates of filling and emptying of the MOT as $\gamma_{trap} \sim 0.1 - 0.01$ Hz, while the escape rate of Rydberg atoms, based on the temperature of the atomic cloud and the size of the trap, is $\gamma_R \sim 10^3$ Hz [5]. Such a ratio of the rates of the processes leads to a high contrast of the obtained two-photon absorption spectrum.

Despite the fact that the radiation field of the MOT beams does not directly affect the population of the Rydberg states, the constant excitation of atoms on the 2S–2P transition during cooling leads to the decay of the coherence between the ground and Rydberg states, which in turn provide an additional line broadening. Calculations show that a good agreement with the experimental data is obtained when the coherence decay rate is qualitatively the same as the excitation rate at the 2S–2P transition by a cooling laser, calculated in the two-level approximation. At minimum total power of the MOT beams γ_{13} is about hundreds of kHz in order of magnitude, which is much higher than the rates of other relaxation processes that can lead to a loss of coherence — γ_3 and γ_R. This suggests that the decay of coherence is determined mainly by the MOT influence. A more detailed investigation of the dependence of the line broadening on the intensity of cooling beams in the MOT is beyond the scope of this paper.

The frequency scanning rate of the UV laser in the experiments under discussion is sufficiently low to consider the system being in a stationary state for any value of detuning δ_2. Therefore, the derivatives on the left-hand side of (1) can be set equal to zero. Since the low saturation of 2S–2P transition the fluorescence signal observed in the experiments is considered to be proportional to the population of the ground state ρ_{11}. Figure 2 shows the calculated shape of the two-photon absorption line in the stationary approximation for one of the experiments. In the experiment the excitation of the 58S level was carried out, the total intensity of the MOT beams was 8.4 mW/cm^2, the intensity of the red laser was 28 mW/cm^2, the intensity of the UV laser was 64 mW/cm^2. Calculation parameters are: $\Omega_1 = 10^8$ s^{-1}, $\Omega_2 = 4 \times 10^5$ s^{-1}, $2\gamma_2 = 37 \times 10^6$ s^{-1}, $2\gamma_3 = 6 \times 10^3$ s^{-1}, $\gamma_{trap} = 2\pi 0.13$ s^{-1}, $\gamma_R = 2\pi 10^3$ s^{-1}, $\gamma_{13} = 3.3 \times 10^6$ s^{-1}. The results of calculations (1) were averaged over the Maxwell distribution for atoms with T=363 μK.

4. Conclusion

The paper presents a model which in a simple three-level approximation makes it possible to describe the results of experiments on two-photon coherent spectroscopy of ultracold Li atoms. The proposed model allows estimating the relative densities of Li atoms in the ground and Rydberg states in these experiments. In this model, both the kinetics of the atoms in the MOT and the broadening of the two-photon absorption line induced by the action of laser cooling beams are taken into account. There is a good agreement of theoretical results with experimental data. As the next step we would like to study the temperature of atoms in MOT under different conditions by using the two-photon spectroscopy with co-propagating and contra-propagating laser beams [6].

References

[1] Ryabtsev I I, Beterov I I, Tret'yakov D B, Entin V M and Yashina E A 2016 *Phys. Usp.* **59** 196
[2] Pillet P and Gallagher T F 2016 *J. Phys. B: At. Mol. Opt. Phys.* **49** 174003
[3] Gallagher T F 2005 *Rydberg Atoms* (Cambridge: Cambridge University Press)
[4] Zelener B B, Saakyan S A, Sautenkov V A, Manykin E A, Zelener B V and Fortov V E 2014 *Jetp Lett.* **100** 366
[5] Sautenkov V A, Saakyan S A, Vilshanskaya E V, Murashkin D A, Zelener B B and Zelener B V 2016 *Laser Phys.* **26** 115701
[6] Sautenkov V A, Saakyan S A, Vilshanskaya E V, Zelener B B and Zelener B V 2017 *Journal of Russian Laser Research* **38** 1
[7] Zelener B B, Saakyan S A, Sautenkov V A, Akulshin A M, Manykin E A, Zelener B V and Fortov V E 2013 *Jetp Lett.* **98** 670
[8] Stenholm S 1984 *Foundations of laser spectroscopy* (New York: John Wiley & Sons, Inc.)
[9] Theodosiou S E 1984 *Phys Rev* E **30** 2881

IOP Conf. Series: Journal of Physics: Conf. Series **945** (2017) 012019 doi:10.1088/1742-6596/945/1/012019

High-precision cutting of polyimide film using femtosecond laser for the application in flexible electronics

D V Ganin, K E Lapshin, A Z Obidin and S K Vartapetov

Physics Instrumentation Centre of Prokhorov General Physics Institute, Russian Academy of Sciences, Troitsk, 142190 Moscow, Russia

E-mail: ganin@optosystems.ru

Abstract. The experimental results of cutting a polyimide film on the optical glass substrate by means of femtosecond lasers are given. Two modes of laser cutting of this film without damages to a glass base are determined. The first is the photo graphitization using a high repetition rate femtosecond laser. The second is ablative, under the effect of femtosecond laser pulses with high energy and low repetition rate. Cutting of semiconductor chips formed on the polyimide film surface is successfully demonstrated.

1. Introduction

Laser fabrication with high-energy femtosecond (fs) lasers has great advantages such as highest precision of micromachining, absence of heat affected zone, ability of 3D direct writing without surface damage, so femtosecond laser has become a powerful tool for high-precision micromachining.

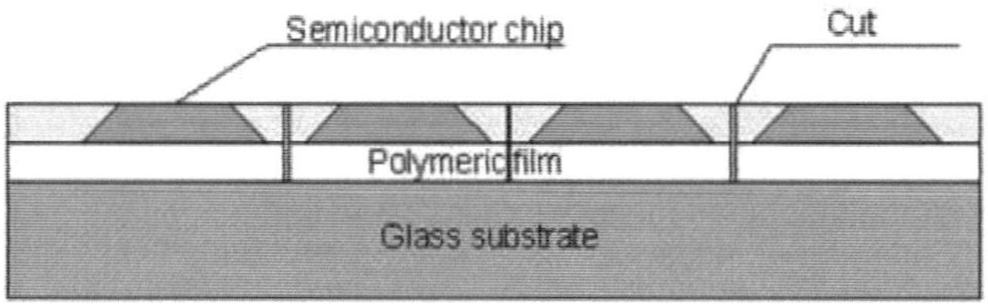

Figure 1. The scheme of laminated layers on the glass substrate.

The development of flexible electronics is one of the main mission of the modern semiconductor industry. Production of amorphous oxide transistors on the surface of thin polymer films opens up wide possibilities for the creation of flexible electronics, in particular, RFID tags, flexible displays and integrated circuits [1]. The most promising technology for manufacturing flexible semiconductor chips and integrated circuits is a method where thin-film IGZO (indium-gallium-zinc oxide) transistors [2] are formed on the polyimide film surface with optical glass substrate (motherglass).

One of the technological problems in the producing of flexible electronics devices is cutting a polyimide film – chip carrier into individual chips after their formation (Figure 1) without glass substrate damage, as it is reusable.

The solution to the good quality polymer chip cutting is to use lasers of ultrashort pulses. No heat-affected zone due to the short pulse duration, high localization of laser exposure due to multiphoton absorption and high pulse repetition rate allows efficiently using femtosecond lasers for polymer film cutting with high performance, while avoiding most of the drawbacks of other technologies. The effect of laser radiation on polymer materials can lead to photochemical conversion to amorphous porous carbon [3–6], and as a result, to a change in optical, electrical and other properties of polymer films, as well as for direct ablation of the material.

In this paper, we demonstrate the possibility of high-precision cutting of a polyimide film with semiconductor integrated circuits formed on its surface, without damaging the glass substrate.

2. Experimental Setup and Samples

Experimental setup is shown on figure 2. We used two different femtosecond laser systems. The first source was a femtosecond Yb regenerative laser system (Avesta Project, RYF-10/35) generating a laser pulses with energies of up to 150 μJ, a wavelength of 1025 nm, pulse duration of 350 fs and a repetition rate from 1 to 2000 Hz. The second source was a femtosecond fiber Yb laser (Optosystems, Fl300) with pulse energies of up to 1 μJ, a wavelength of 1053 nm, pulse duration of 350 fs and a repetition rate of 2 MHz.

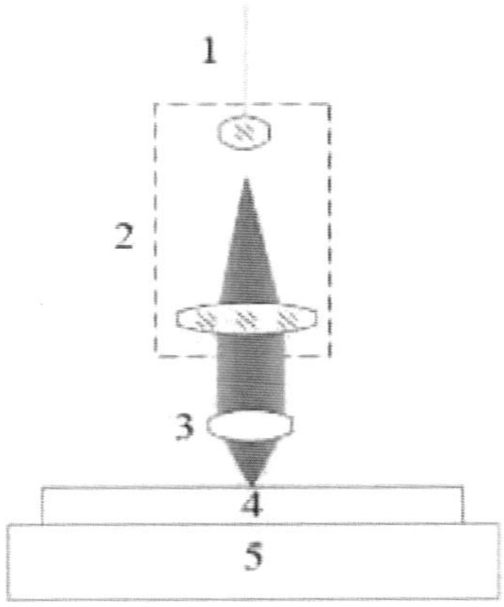

Figure 2. Scheme of experimental setup: (1) femtosecond laser beam; (2) beam expander; (3) focusing lens on Z-stage; (4) sample; (5) XY translation stage.

To control the power and the pulse energy of the linear- polarized laser beam, we used a rotatable half-wave plate and thin film polarizer. Beam expander with 4x magnification was used to expand laser beam. A lens of 54-18-23-1064 nm, Special Optics with a working distance (WD) of 23 mm, N.A. = 0.39 was used as a focusing system. The lens was mounted on Z-translator to control the radiation focal depth. Radiation was focused into the sample mounted on XY translator. Pictures of exposure of fs laser pulses were obtained using an optical microscope (Nikon LV100D) in transmitted light in phase-contrast mode, which allows you to see the presence/absence of damage to the surface of the glass substrate.

3. Cutting Flexible Electronic Chips with Ultrashort Pulse Lasers

Our experiments have shown that cutting of polyimide film – chip carrier can be successfully performed by femtosecond lasers without damaging the glass substrate. The use of two FS lasers with different energy-repetition rate characteristics has shown two fundamentally different film cutting processes – photochemical and ablative.

3.1. Ablative Cutting

The use of the first laser source with a pulse energy reduced to 2 μJ results "ablative" mode, when the material is ablated from the zone of exposure (Figure 3) to femtosecond laser pulses. The bulk of the material is removed during ablation, and the surface surrounding the cut is significantly contaminated. The graphite (soot) layer on the cutting edges also worth noting. No melting of the area surrounding the cut is observed. The following parameters were used: scanning speed 3 mm / sec, laser pulses were focused under the surface of the film at a depth of 40 μm.

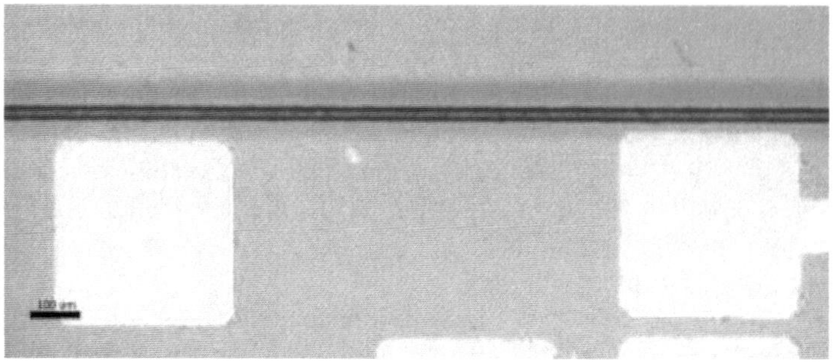

Figure 3. Photo of the cut between chips. Ablative mode. E_{pulse} = 2 μJ, NA = 0.39, scanning rate = 3 mm/sec. Cut width about 20 μm.

3.2. Photochemical Cutting

Fundamentally different cutting mode is implemented when using a fs laser with high repetition rate (1 MHz) and low pulse energy <120 nJ, where the area of exposure to FS pulses undergoes photochemical transformations (Figure 4). There are photochemical transformations of the polyimide film – "soot" is formed in the area of exposure to fs pulses. This soot is very brittle and can be easily removed by mechanical cleaning or flushing. Photochemical cutting mode can be high performance way for cutting polyimide film-carrier in flexible electronics.

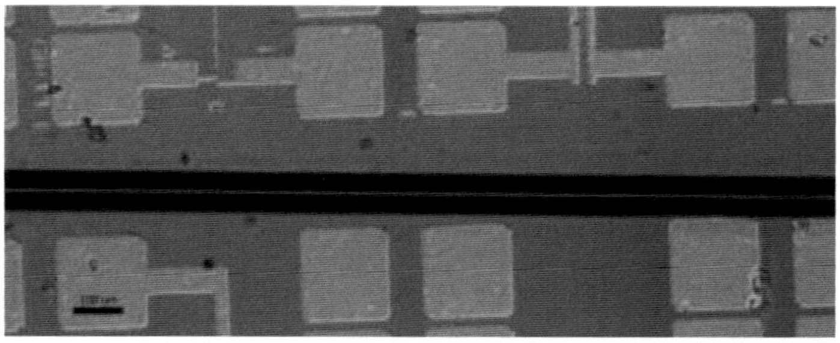

Figure 4. Photo of the cut between chips. Photochemical mode. Frequency – 2 MHz and E_{pulse} = 120 nJ. Cutting speed – 15 mm/sec. Cut width about 80 μm.

4. Discussion of the Results

In "ablative" cutting mode, it is possible to provide a uniform cut with sharp edges. In the above experiments, the degree of overlap between adjacent focal spots during the scanning process was 25%. Using the focusing lens with NA = 0.39 allowed to reach the cutting width of 20 μm. Relatively high pulse energies have a significant probability of damage to the glass substrate without precise radiation focusing in the desired area. Surface damage or change in the glass refractive index was observed in experiments when focusing pulses to a depth greater than 70 μm.

"Ablative" cutting mode using a standard femtosecond fibre laser (1 MHz, 2 μJ, 350 fs) allows achieving cutting speeds of 250 mm/sec. The speed may be increased when using more efficient laser, provided no strong influence of thermal accumulation effects. Additional studies on selection of the optimal scanning parameters, along with the optical system optimization will reduce the cutting width, improve its quality and increase efficiency.

"Photochemical" cutting mode has its own features. Low pulse energies below the glass damage threshold, do not leave traces of impact on the glass substrate regardless of focusing precision. This cutting mode is realized in the heat accumulation mode. In our opinion, photochemical decomposition of the polyimide film occurs at high temperatures, and a graphitized substance formed has a noticeable electrical conductivity. Specific electrical conductivity of the graphitized line measured by the method similar to the one proposed in [4] was about 10^4 S/m, which is consistent with the data of [7] on the polyimide film graphitization by heating.

Possible surface contamination with degradation products is an undesirable effect. Also it is worth noting that typical cutting width is about 100 μm. We think such a high cutting width is due to photochemical and thermal processes as a result of heat accumulation and diffusion. At the same time, there is a range of parameters where the cutting width of less than 50 μm and the cutting speed of 50 mm/sec can be achieved.

Direct laser graphitization of the polyimide film can be applied in electronics and terahertz optics [7,8].

5. Conclusion

The use of two fs laser sources with different energy-repetition rate characteristics showed the presence of two fundamentally different polyimide film cutting methods. Both methods have their advantages and drawbacks. The advantages of "ablative" cutting include good cutting quality, no large contamination and possibility to change the cutting width. At the same time there are some drawbacks, in particular possible damage to the substrate without high-precision focusing, and the need of high-energy fs laser source.

The advantages of "photochemical" cutting include technology simplicity, low fs laser pulse energy requirements, high performance, no damages to the substrate. The drawbacks include film surface contamination and relatively high cut width.

Acknowledgements

This work was partially supported by Foundation for Assistance to Small Innovative Enterprises in Science and Technology, program U.M.N.I.K. (10624GU/2016 No. 0023590), and the Competitiveness Program of National Research Nuclear University MEPhI.

References

[1] Nomura K, Takagi A, Kamiya T, Ohta H, Hirano M and Hosono H 2006 Amorphous oxide semiconductors for high-performance flexible thin-film transistors *Japanese J. Appl.Physics, Part 1 Regul. Pap. Short Notes Rev. Pap.* **45** 4303–8

[2] Nag M, Bhoolokam A, Smout S, Willegems M, Muller R, Myny K, Schols S, Ameys M, Genoe J, Ke T H, Vicca P, Ellis T, Cobb B, Kumar A, Steen J-L P J van der, Gelinck G, Fukui Y, Obata K, Groeseneken G, Heremans P and Steudel S 2015 Circuits and AMOLED display with self-aligned a-IGZO TFTs on polyimide foil *J. Soc. Inf. Disp.* **22** 509–17

[3] Least B T and Willis D A 2013 Modification of polyimide wetting properties by laser ablated conical microstructures *Appl. Surf. Sci.* **273** 1–11

[4] Morita N, Shimotsuma Y, Nishi M, Sakakura M, Miura K and Hirao K 2014 Direct microcarbonization inside polymer using focused femtosecond laser pulses *Appl. Phys. Lett.* **105**

[5] Cai J, Lv C and Watanabe A 2016 Cost-effective fabrication of high-performance flexible allsolid-state carbon micro-supercapacitors by blue-violet laser direct writing and further surface treatment *J. Mater. Chem. A* **4** 1671–9

[6] Gao W, Singh N, Song L, Liu Z, Reddy A L M, Ci L, Vajtai R, Zhang Q, Wei B and Ajayan P M 2011 Direct laser writing of micro-supercapacitors on hydrated graphite oxide films. *Nat. Nanotechnol.* **6** 496–500

[7] Venkatachalam S, Bertin D, Ducournau G, Lampin J F and Hourlier D 2016 Kapton-derived carbon as efficient terahertz absorbers *Carbon N. Y.* **100** 158–64

[8] In J Bin, Hsia B, Yoo J H, Hyun S, Carraro C, Maboudian R and Grigoropoulos C P 2015 Facile fabrication of flexible all solid-state micro-supercapacitor by direct laser writing of porous carbon in polyimide *Carbon N. Y.* **83** 144–51

IOP Conf. Series: Journal of Physics: Conf. Series **945** (2017) 012020 doi:10.1088/1742-6596/945/1/012020

Results of modeling of thermal-hydraulic processes in the sodium cooled experimental assembly in approach of an anisotropic porous body

M V Bayaskhalanov, M N Vlasov, A S Korsun, I G Merinov and V V Silvestrov

National Research Nuclear University MEPhI (Moscow Engineering Physics Institute), Kashirskoe shosse 31, Moscow, Russia

E-mail: mr.bayashalanov@mail.ru

Abstract. Results on the second stage of the program APMod module testing are presented. As well as at the first stage of testing good high-quality and quantitative coincidence of hydrodynamic characteristics of the sodium coolant in the area behind blockage has been received. At the same time the noticeable divergence of settlement and experimental fields of temperature remained. The analysis of the possible reasons of a results divergence of calculation with experimental data is carried out.

1. Introduction

Modeling of thermal-hydraulic processes in the core of the modern fast reactors made from the jacketless fuel assemblies and cooled by the liquid metal coolant demands use of three-dimensional mathematical models. Use of an anisotropic porous media approach [1-3] allows to solve an objective with rather low expenses of the computer resources (in comparison with CFD codes).

The purpose of this work was testing of the program APMod module developed for the modeling of heat and mass transfer processes in perspective nuclear reactors cores and heat exchange equipment in approach of an anisotropic porous media. Testing was carried out during the simulation of the sodium coolant flow in experimental fuel assembly with partial blockage of the flow section.

2. Experimental installation

The results received on the experimental KNS-Test [4] installation (Figure 1) used for a research of processes of a heat and mass transfer in a core of the SNR-300 reactor were used as an experimental data. Assembly contained 169 smooth cylindrical rods, some of which was heated by electric current. Sodium coolant flow simulation in assembly was carried out for a case of partial flow section blockage at 21% overlapping of an angular part of cross section of assembly (Figure 1). Blocking settled down in 40 mm from the beginning of a zone of energy release. As boundary conditions were set coolant velocity on an entrance to assembly – 4 m/s, his temperature – 668°C and pressure at the exit (it was accepted equal atmospheric).

The first stage of testing has been described in work [5]. The results of the first stage showed operability of the APMod module and both good quantitative and high-quality coincidence of settlement and experimental characteristics for the sodium coolant flow. However, the divergence of the coolant temperature fields was observed. Therefore the main attention was paid to establishment of the reasons of a results divergence of calculation and an experiment in definition of the field of temperature behind blockage and optimization of the used settlement grid.

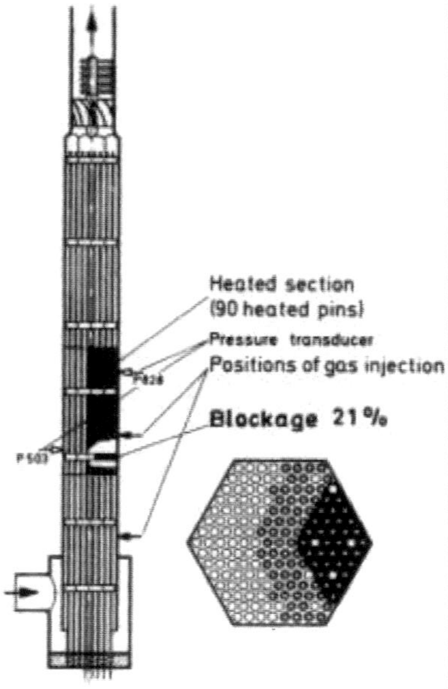

Figure 1. Experimental KNS-Test installation

3. Calculation results

Optimization of a settlement grid has allowed to reduce number of elements of splitting from 106470 to 3888 elements (Figure 2) that has sharply reduced time of calculation performance at preservation of accuracy of the received results.

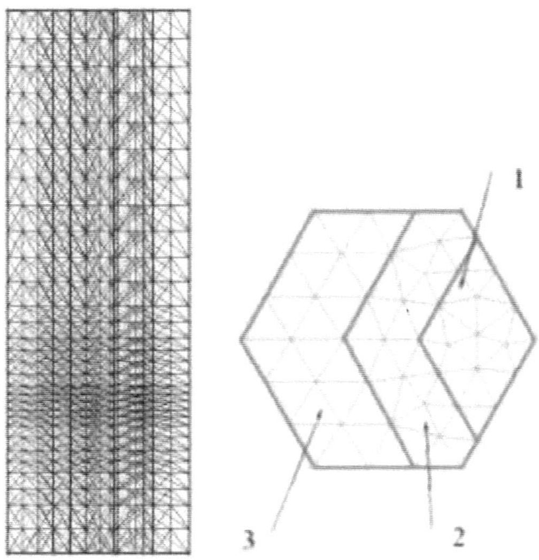

Figure 2. Settlement grid used for computer simulation: 1 – region with blockage and heat release, 2 – region with heat release, 3 – region without flow blockage and heat release

The resulting calculated coolant velocity distribution in longitudinal section of assembly is shown on Figure 3 with the characteristic points noted. Received distribution is qualitatively coinciding with the experimental results. In the coolant flow behind of blockage a sector of a toroidal vortex was watched. For the analysis of the reasons leading to a divergence of settlement and experimental data there was a carried-out series of calculations, with various coefficients of blockage resistance. Results of comparison of hydrodynamic characteristics of these calculations with experimental are given in Table 1.

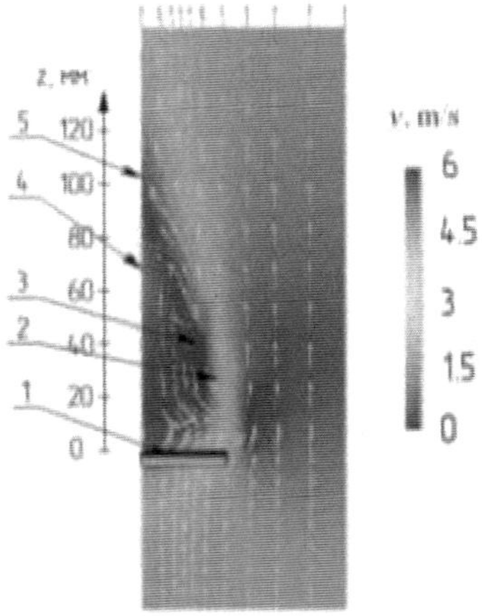

Figure 3. Velocity distribution in longitudinal section of assembly (1 – blockage; 2 – mixing zone between main flow and recirculating flow; 3 – center of the vortex; 4 – reverse flow; 5 – upper stagnation point)

Table 1. Coolant flow hydrodynamic parameters

Parameter	Experiment	Calculation (blockage resistance 4500,1/s)	Calculation (blockage resistance 6000, 1/s)	Calculation (blockage resistance 9000, 1/s)
Height of the stagnant zone, mm	105	87	97	120
Vortex center position, mm	45	46	46	45
Relative velocity of reverse flow	0.17	0.08	0.14	0.18

As well as at the first stage of testing good high-quality and quantitative coincidence of hydrodynamic characteristics of a sodium coolant flow in the area behind blockage has been received. It is even observed at less detailed grid.

For comparison of settlement fields of temperatures with experimental, the received values of temperature were normalized by the technique used in work [4]. The short description of this technique is provided in work [5]. In the Figure 4 results of comparison of radial fields of temperatures in the area behind blockage at various coefficients of blockage resistance are shown.

As it follows from figure 4, calculation results more consistent with experimental data at increasing of resistance in flow blockage region. The remaining noticeable deviation of settlement dependences on experimental curves in the area over the blockage adjacent to an assembly jacket (at big x) at long distances behind the blockage can be explained by heat exchange between assembly and environment in the experiment that wasn't considered in calculations.

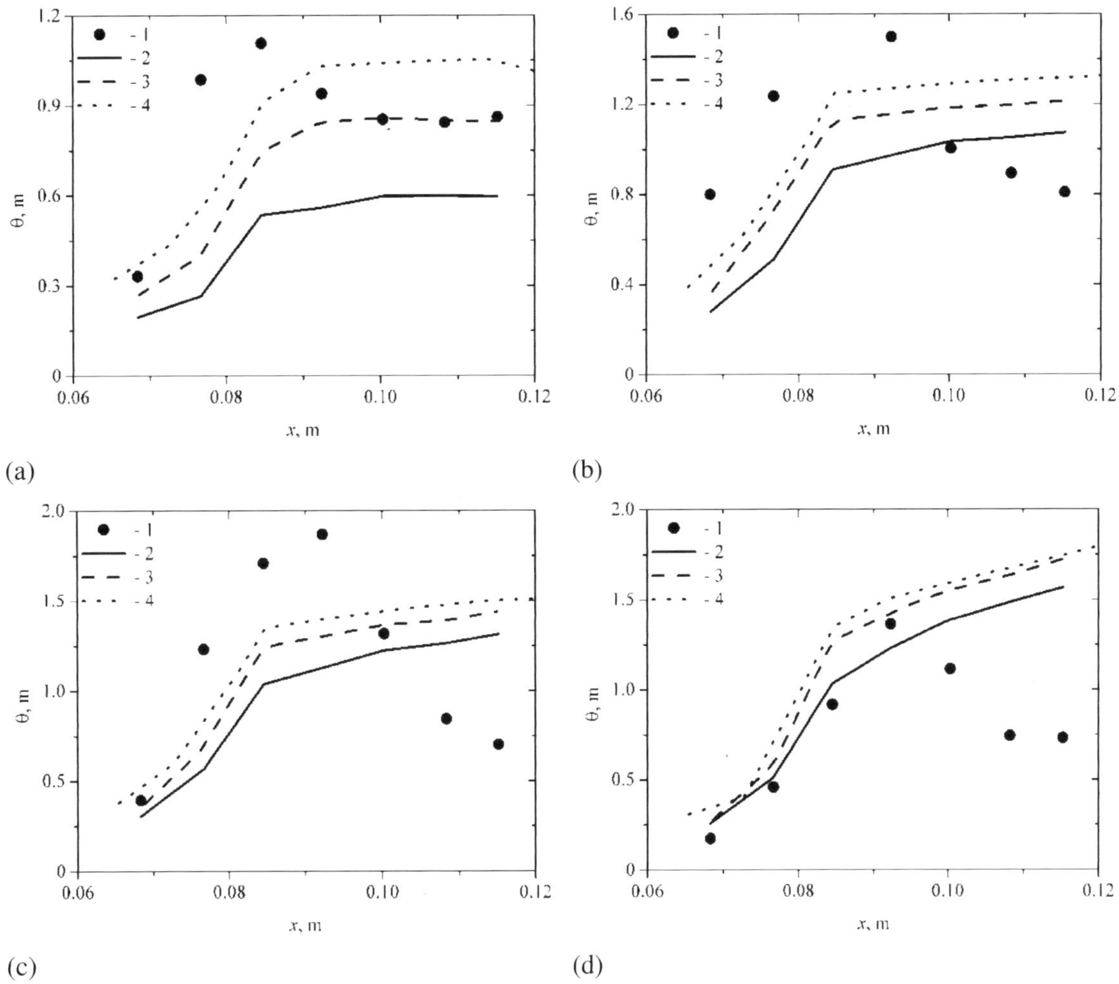

Figure 4. Radial temperature distributions at different distances from the blockage (a – 10 mm, b – 30 mm, c – 50 mm, d – 70 mm): 1 – experimental data [4], 2 – calculation (blockage resistance 4500, 1/s), 3 – calculation (blockage resistance 6000,1/s), 4 – calculation (blockage resistance 9000,1/s)

4. Conclusion

The optimization of a settlement grid which has allowed to increase significantly the speed of calculations without loss of accuracy has been made. The executed calculations have shown that the main reason for a divergence of settlement and experimental fields of temperatures is the description of blockage in the program module by a high hydraulic blockage resistance. The blockage remains partially permeable, unlike impenetrable blockage in an experiment. Justice of this conclusion is confirmed by the fact that increase in blockage resistance leads to improvement of coincidence of results of calculation to an experiment. In general the conducted calculations confirm operability of the program APMod module.

Acknowledgements

This work was supported by Competitiveness Program of National Research Nuclear University MEPhI.

References

[1] Korsun A C, Maslov Yu A, Merinov I G et al. 2013 *Nuclear Physics and Engineering* **4** (7) 619 (DOI: 10.1134 / S2079562913070063) (In Russian).
[2] Korsun A S, Kruglov V B, Merinov I G et al. 2014 *Problems of Atomic Science and Technology, Series: Nuclear and Reactor Constants* **No 2** 87 (In Russian).
[3] Vlasov M N, Korsun A S, Maslov Yu A et al. 2016 *Thermophysics and Aeromechanics* **23** (2) 201 (DOI: 10.1134/S0869864316020062).
[4] Huber F, Peppler W, Report No 3927 (Karlsruhe, 1985).
[5] Bayaskhalanov M V, Merinov I G, Korsun A S and Vlasov M N 2017 *Journal of Physics: Conference Series*, **781**(1) 1 (DOI: 10.1088/1742-6596/781/1/012049).

IOP Conf. Series: Journal of Physics: Conf. Series **945** (2017) 012021 doi:10.1088/1742-6596/945/1/012021

Dependence of critical heat flux on the rate of increase in heat supply

M.I. Delov, K.V. Kutsenko, D.M. Kuzmenkov, A.A. Lavrukhin, S. Laouar, Yu.E. Litvintsova, A.S. Marchenko, Yu.A. Maslov

National Research Nuclear University MEPhI (Moscow Engineering Physics Institute), Moscow, Russia

E-mail: dmitrii.kuzmenkov@mail.ru

Abstract. The paper presents a generalized analysis of the influence of the rate of increase in thermal load and pressure on the critical heat flux in liquid nitrogen and other cryogenic media under conditions of unsteady heat supply. The results of the investigation can be used in the calculation of high-precision equipment based on high-temperature superconductors (HTSCs), such as current limiters, solenoids, etc.

1. Introduction

Investigation of the regularities of the non-stationary heat transfer crisis [1] is of interest for understanding physics of boiling fluids. It is shown in [2, 3] that due to a sharp increase in the heat release in the heater from a zero level the bubble boiling crisis occurs for liquid nitrogen and some other liquids when heat fluxes $q_{cr,\ min}$ are much lower than the value of the first critical heat flux [3] obtained by a quasi-stationary (slow) increase in the density of the heat flux q. This leads us to conclude that the magnitude of the critical heat flux depends significantly on the rate of increase in heat release ω. It was obtained experimentally in [4] that for a stepwise increase in the capacity of heat release there is a boundary pressure p_{bor} above which $q_{cr.\ min} = q_{cr2}$, where q_{cr2} is the minimum value of the heat flux at which a stable film boiling can exist. Particular scientific interest is the dependence of $q_{cr,\ min}$ not only on the fluid pressure and on the heater geometry but also on the rate of increase in its heat release. The study of this dependence is also of practical value for the transition of HTSC devices to the resistive state, which is accompanied by a significant change in the heat release. Moreover, when the heat flux $q_{cr,\ min}$ is reached on the surface of the superconductor, the heat transfer to the coolant deteriorates significantly, which can lead to its irreversible destruction. This paper presents a generalized analysis of the influence of the rate of increase in the heat release, pressure, and heater diameter d on $q_{cr,\ min}$.

2. Theoretical study

Transition to film boiling with non-stationary heat supply occurs during a certain time interval τ_{cr}, that is conventionally referred to as "critical". In this respect, the critical time interval for some fluids (nitrogen, oxygen, neon, etc.) is determined by the time before boiling τ_{eb} begins [4]. In the first approximation, in case of such fluids the necessary condition for film boiling with a rapid change in capacity is the equality of superheating of heat-releasing surface to the limiting overheating of homogeneous nucleation ΔT_{lim} [5].

If free convection develops near the heat-release surface after completing the stage of non-stationary heat conduction, the transition to a film boiling regime does not occur. If this is not the case, the stage of metastable boiling will end with a transition to a film boiling regime and a heat transfer crisis. Thus, the criterion for the crisis onset is the equality of time before the boiling point of the fluid τ_{eb} and the characteristic time for the development of convection τ_{conv} when the wall ΔT_{lim} is overheated. The characteristic time for the development of convection is determined through reaching the critical Rayleigh number in the thermal boundary layer.

In [4] expressions were derived for $q_{cr,\,min}$ for the case of a stepwise change in the density of the heat flux on the surface of the heater:

$$q_{cr,\,min} = \frac{2}{\mathrm{Ra}_{cr}^{1/3}} \left(\frac{\lambda^3 \cdot \Delta T_{lim}^4 \cdot g \cdot \beta}{\nu \cdot a} \right)^{1/3}, \qquad (1)$$

where Ra_{cr} is the critical Rayleigh number (for horizontal heaters $\mathrm{Ra}_{cr} \approx 500$), λ, β, ν are the coefficients of thermal conductivity, thermal expansion, and kinematic viscosity of the liquid, respectively, and g is acceleration of gravity.

The rate of increase in the heat load $\omega = \left\langle \frac{dq}{d\tau} \right\rangle$ can be different. Stepwise increase in capacity ($\omega \to \infty$) and quasistationary heat supply ($\omega \to 0$) are the two limiting cases of variation of $q(\tau)$. In the case of quasistationary heat supply, the minimum value of the nonstationary critical heat flux $q_{cr,\,min}$ coincides with q_{cr1}. In case of a stepwise increase in capacity, the crisis occurs $q_{cr,\,min} < q_{cr1}$. Applying the results of work [4] to the heat release law in the form $q = C{\cdot}\tau^m$, we considered the dependence of the critical heat flux on the rate of increase of the heat load. As a result of the calculation, an expression for the limiting rate of capacity increase ω_{bor} was derived, where the transition to film boiling at heat fluxes $q < q_{cr1}$ is yet possible in the case of non-stationary heat supply:

$$\omega_{bor}(m) = \frac{2}{m+1} \cdot \frac{6 \cdot \lambda \cdot g \cdot \beta \cdot \Delta T_{lim}^2}{\nu \cdot \mathrm{Ra}_{cr}}, \qquad (2)$$

Figure 1 shows the influence of the rate of increase in the thermal load ω on the critical heat flux. It follows from the figure that for $\omega < \omega_{bor}$ the boiling crisis will occur with flows equal to q_{cr1}. With exponent m increasing, the boundary value ω_{bor} decreases, and for $m \to \infty$ the boundary value of the rate of increase in the thermal load $\omega_{bor} \to 0$, which corresponds to the case of a stepwise increase in capacity. It is important to note that the value of $q_{cr,\,min}$ does not depend on the pattern of the increase in heat release and is described by formula (1) for any m.

Figure 2 shows a graph of the dependence of $q_{cr,\,min}$ on the pressure in liquid nitrogen for a heater with a diameter of 100 μm in comparison with the experimental data in [4]. As can be seen from the graph, at pressures greater than $p_{bor} \approx 2$атм, the calculated value $q_{cr,\,min}$ becomes less than q_{cr2}, and there exists no stable film boiling on the heater. Therefore, at $p > p_{bor}$, the value of the minimal non-stationary critical heat flux will coincide with the value of q_{cr2}. According to our calculations, the value of p_{bor} only depends on the heater geometry if its diameters are smaller than those of the Laplace constant (Figure 3).

Figure 4 shows a dependence graph of the boundary rate of increase, related to the critical heat flux, on the reduced pressure for various cryogenic fluids in the case of the exponential function of the heat release function at $m = 1$. As can be seen from the figure, the above dependences for different fluids are similar to each other.

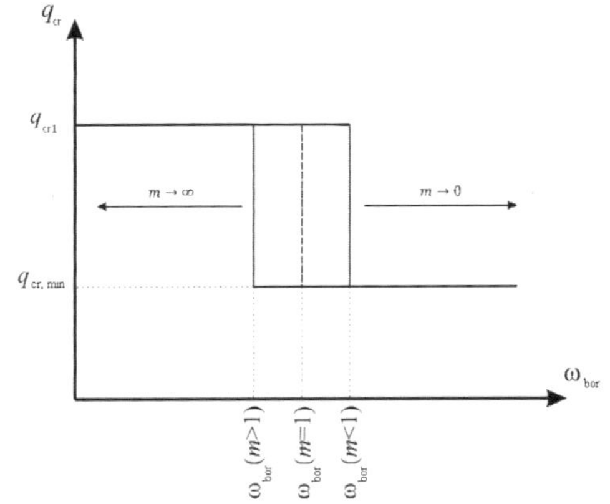

Figure 1. Schematic dependence of the non-stationary critical heat flux on the increase rate of the heat load for various types of heat release functions $q=C\cdot\tau^m$.

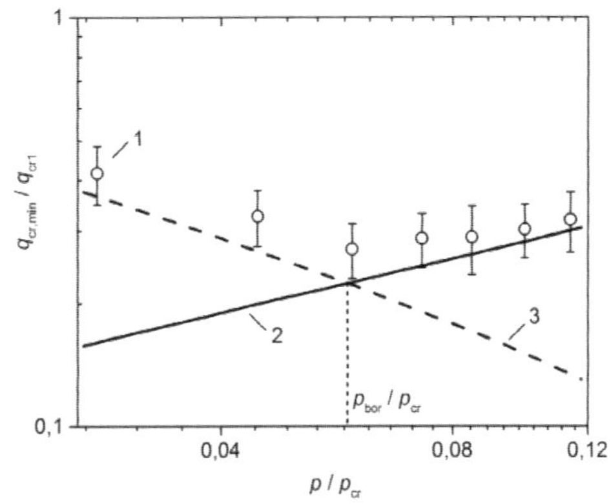

Figure 2. Critical heat flux as a function of reduced pressure [4]: 1 – experimental data, 2 - second critical heat flux, 3 – minimum critical heat flux.

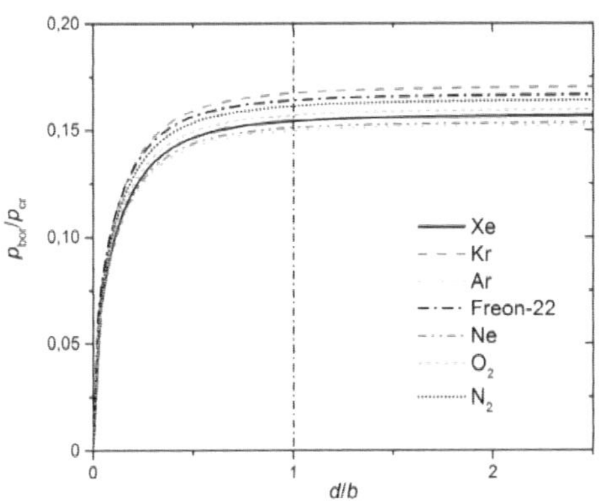

Figure 3. Extremum reduced pressure as function of heater geometry for different cryogenic fluids [4].

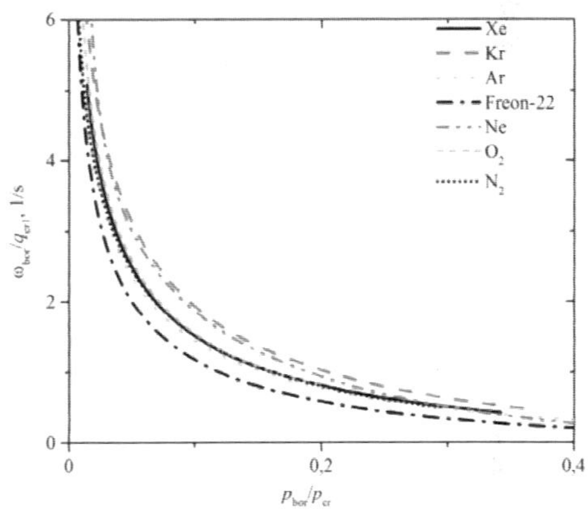

Figure 4. Dependence of the boundary increase rate on the reduced pressure.

Thus, in order to calculate the critical heat flux, the following relation can be offered:

$$q_{cr}(\omega, p, d, m) = \begin{cases} q_{cr1}, & \omega < \omega_{bor}(p,m) \\ q_{cr,\,min}, & \omega \geq \omega_{bor}(p,m),\ p < p_{bor}(d) \\ q_{cr2}, & \omega \geq \omega_{bor}(p,m),\ p \geq p_{bor}(d) \end{cases} , \qquad (3)$$

where $q_{cr,\,min}$ and ω_{bor} are calculated using formulas (1) and (2), respectively. The value of the reduced border pressure for planar and cylindrical heaters with a diameter larger than the Laplace constant is $p_{bor} \approx 0{,}16 \cdot p_{cr}$ for all the fluids considered.

3. Conclusion

The paper has put forward a universal method for calculating non-stationary critical heat flux, depending on the rate and pattern of the increase in heat release, heater geometry and pressure for various media. Nevertheless, it should be noted that this model requires some additional experimental confirmation and refinement for cylindrical heaters of a smaller diameter. A detailed study of the dependence of the critical Rayleigh number on the diameter and orientation of the heater in the field of gravity is also of particular interest.

Acknowledgment
This research was supported by a grant of the Russian Science Foundation (project №16-19-10548).

References
[1] Kutateladze S S. 1961 Boiling heat transfer *Int. Journal of Heat and Mass Transfer* **4** 31-45
[2] Pavlenko A N and Chekhovich V Yu 1986 Burn-out heat transfer in unsteady state heat generation in a cryogenic fluid *Heat transfer. Soviet research* **18** (3) 83-94
[3] Deev V I, Kharitonov V S, Kutsenko K V and Lavrukhin A A 2004 Transient boiling crisis of cryogenic liquids *International Journal of Heat and Mass Transfer* **47**/25 5477—82
[4] Balakin B V, Delov M I et al. 2017 Boiling crisis in cryogenic fluids during unsteady heat supply *International Journal of Heat and Mass Transfer* **111** 1107-11
[5] Sinha D N, Semura J S and Brodie L C 1982 Homogeneous nucleation in ^{4}He: A corresponding-states analysis *Phys. Rev* A 26 (3) 1048-61

The relationship between dynamic and average flow rates of the coolant in the channels of complex shape

V N Fedoseev[1], M I Pisarevsky[2] and Y N Balberkina[3]

[1]Docent, National Research Nuclear University "Moscow Engeneering Physics Institute, Moscow, Russia
[2]Engeneer, National Research Nuclear University "Moscow Engeneering Physics Institute, Moscow, Russia
[3]Student, National Research Nuclear University "Moscow Engeneering Physics Institute, Moscow, Russia

E-mail: MIPisarevskij@mephi.ru

Abstract. This paper presents interconnection of dynamic and average flow rates of the coolant in a channel of complex geometry that is a basis for a generalization model of experimental data on heat transfer in various porous structures. Formulas for calculation of heat transfer of fuel rods in transversal fluid flow are acquired with the use of the abovementioned model. It is shown that the model describes a marginal case of separated flows in twisting channels where coolant constantly changes its flow direction and mixes in the communicating channels with large intensity. Dynamic speed is suggested to be identified by power for pumping. The coefficient of proportionality in general case depends on the geometry of the channel and the Reynolds number (Re). A calculation formula of the coefficient of proportionality for the narrow line rod packages is provided. The paper presents a comparison of experimental data and calculated values, which shows usability of the suggested models and calculation formulas.

1. Introduction

The determination of the hydrodynamic resistance and heat transfer is the primary goal of considering the convective transfer in the heat exchanger. The heat transfer coefficient of rod assemblies depends on large list of parameters: the longitudinal and transversal steps of package, hydraulic diameter, flow velocity, flow properties and etc.

$$Nu = f(\mathrm{Re}, \mathrm{Pr}, {}^{s_1}\!/_{d}, {}^{s_2}\!/_{d} \ldots) \tag{1}$$

Since the parameters in equation (1) are interconnected, a determination of this parameters separate and comprehensive influence on heat transfer becomes crucial objective. The correct choice of the equation structure could arise after numerical analysis of experimental data.

2. Construction of the model

Two factors have served as a basis for making new generalizations. The first factor is the construction of a unified physical model to summarize information about any structures. The second one is a good agreement of the calculations, that is performed with the use of the recommended empirical formulas [1-7], with experimental data in in-line bundles of tubes with a large value of porosity ($\Pi \geq 0.3$). The difference in the experimental data and calculated values does not usually exceed 25 - 30 %. However, the recommended formulas are not appropriate for high-density line bundles. The difference in the experimental values and calculated data varies from 50 to 300 % (Fig.1). The high density bundles are supposed to be the ones with no more than 30 % porosity value ($\Pi < 0.3$).

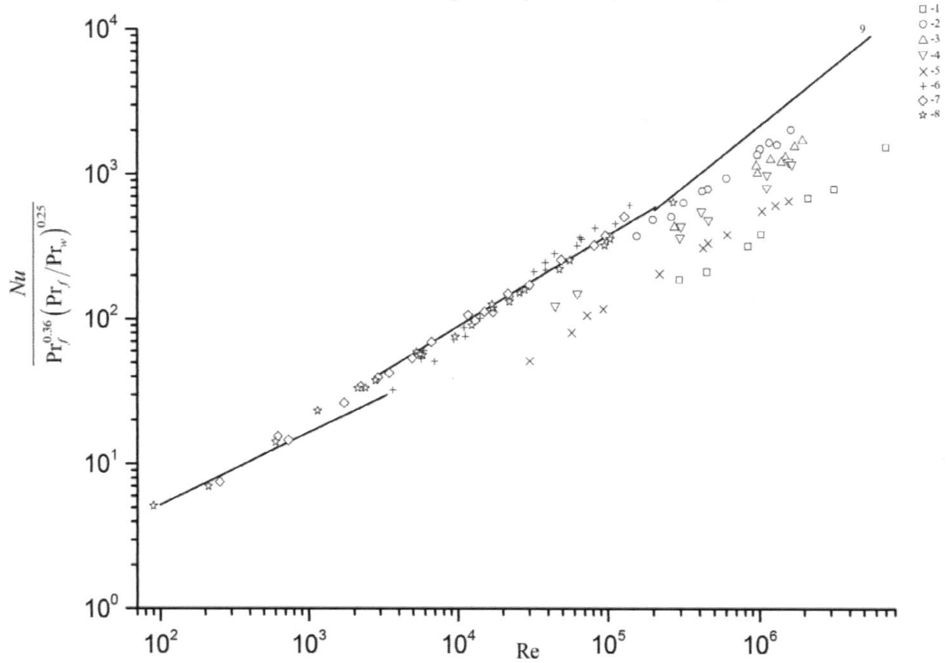

Figure 1. Dependence of heat transfer complex on Re number for line bundles: : 1- $1{,}053\times1{,}053$, 1- $1{,}026\times1{,}026$, 3- $1{,}020\times1{,}020$, 4- $1{,}013\times1{,}013$, 5- $1{,}008\times1{,}008$, 6- $2{,}63\times1{,}10$, 7- $2{,}00\times2{,}00$, 8- $1{,}65\times2{,}00$, 9 – experimental values obtained by A.A. Zukauskas [1].

In [6] the next dependence of heat transfer resistance in turbulent flow in smooth channels, which is construvted with the use of factor analogy, is recommended:

$$\frac{c_f/2}{St} = k_1 + k_2\sqrt{\frac{c_f}{2}}(\mathrm{Pr}^{2/3} - 1) \tag{2}$$

where $St = \dfrac{Nu}{\mathrm{Re}\,\mathrm{Pr}}$ is the Stanton number, $c_f = \xi/4$ – friction coefficient, ξ – coefficient of hydraulic resistance, $k_1 = 0.93$, $k_2 = 12.5$. Assuming $k_1 = 1$, $k = k_2 = 12.7$, $n = \dfrac{2}{3}$, a general equation is obtained:

$$St = \frac{\xi/8}{1 + k\sqrt{\xi/8}\,(\mathrm{Pr}^n - 1)} \tag{3}$$

In [10-11], the universality of equation (3) and its applicability to the calculation of average heat transfer in channels of complex shape with tear currents is assumed. The main task is to find the relationship between the dynamic and average coolant flow rates in the channel.

3. Interconnection of dynamic and average flow rates of the coolant

An important characteristic of coolant flow rate distribution near the wall of pipe is the dynamic flow rate v_* that is associated with the average speed $\overline{V}$ and the coefficient of resistance ξ by the equation:

$$v_* = \bar{v}\sqrt{\frac{\xi}{8}} = \bar{v}\sqrt{\frac{c_f}{2}}$$

Considering this, the following equation appears:

$$\frac{\bar{v}}{v_*} = \sqrt{\frac{8}{\zeta}} = f(\mathrm{Re}_*)$$

The dynamic flow rate is supposed to be determined through the cost of coolant pumping power. Then, with the consideration of dimensions:

$$v_* = c\left(\frac{v\upsilon\nabla P}{\rho}\right)^{0.25} \tag{4}$$

where υ - the kinematic viscosity coefficient, C – coefficient of proportionality, which depends in the General case on the geometry of the channel and Re number.

To determine the value of the constant «C» it is necessary to define function $f(\mathrm{Re}_*)$. Previously, this function was obtained for the pipe (the first equation of the following system) and the plate (second equation of the following system).

$$f(\mathrm{Re}_*) = \frac{\bar{v}}{v_*} = \sqrt{\frac{8}{\xi}} = \begin{cases} 2.5\ln \mathrm{Re}_* \\ 6.33\mathrm{Re}_*^{\frac{1}{8}} \end{cases} \tag{5}$$

The obtained values give similar results after substitution of both functions in equation (5). The difference is less than 10%. In further generalization function $f(\mathrm{Re}_*)$ was used for the plate.

4. Data generalization

The calculation of the coefficient of proportionality «C» for a dense line bundles is implemented. It is shown that «C» does not depend on the Re number for each dense packing (figure2).

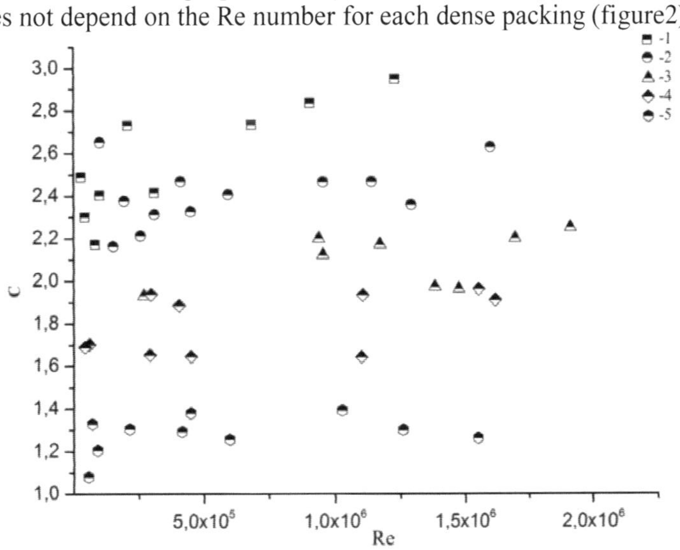

Figure 2. The dependence of coefficient «C» from Re number for narrow line bundles ($\Pi < 0.3$):1- 1.053×1.053, 2- 1.026×1.026, 3- 1.020×1.020, 4- 1.013×1.013, 5- 1.008×1.008.

The formula (6) is recommended for calculations since the usage of this equation in (3) outputs good agreement with experimental data [1]. The deviation of calculated values from the experimental data does not exceed 15%.

$$c = -45.6 + 350.5\Pi - 635.9\Pi^2 \tag{6}$$

5. Conclusion

The paper presented interconnection of dynamic and average flow rates of the coolant and experimental data on heat transfer in various porous structures generalization model, which has several advantages. The first one is in fact that only one empirical constant presents in generalization. Additionally, new calculation formula is obtained, which has more accurate agreement with the experimental data [1].

Acknowledgments

The study was performed by the Russian Science Foundation (project No. 16-19-10548).

References

[1] Zukauskas A A, Makaryavichus V and Schlanchyauskas A 1968 *«Mintis» Vilnius* Heat transfer of tubes bundles in transverse fluid flow p 192 (In russian)
[2] Migay V K and Firsova E V 1986 L. *Nauka* Heat transfer and hydraulic resistance of tubes bundles p 195 (In russian)
[3] Pogorelov A I 1999 *Odessa: Chernomorye* Heat and mass exchange (Basis of theory and calculation) p 128 (In russian)
[4] Andreev L P , Nikulshin V R and Andruschenko A M 2006 issue 1(25) *Odessa Polytechnic university* Generalized equation of heat transfer during crossflow of bundles of smooth tubes (In russian)
[5] Pis'mennyi E N 2010 *Thermal Engineering* A Universal Relation for Calculating Convective Heat Transfer in Transversely Streamlined Bundles of Smooth Tubes Vol 57, No 3, p 219–226
[6] Zukauskas A 1982 Science Convective transfer in heat exchangers p 470 (In russian)
[7] Thermal calculation of boilers (Normative method) 1998 *SPb NPO CKTI* p 256 (In russian)
[8] Petukhov B S *et al* 1986. Heat transfer in nuclear power plants. Energoatomizdat p 233 (In russian)
[9] Petukhov B S and Kirillov V V 1958 *Thermal Engineering* To the question of heat transfer in a turbulent flow of fluid in the tubes №4 (In russian)
[10] Kokorev L S et al 1987 High Temperature On the relationship between flow resistance and heat transfer in porous media 6 92 (In russian)
[11] Fedoseyev V N *et al* 1987 *Thermal Engineering* The universal dependence of heat transfer from the pressure gradient in porous mediums p 61-64 (In russian)

Investigation of MIS-sensor sensitivity to vapor of unsymmetrical dimethylgydrazine in air

D V Filipchuk[1,2], A V Litvinov[1], M O Etrekova[1,2] and D A Nozdrya[2]

[1]Department of Solid State and Nanosystems, National Research Nuclear University "MEPhI", 115409, Russia, Moscow, Kashirskoe highway, 31
[2]OOO NPF "INKRAM", 109341, Russia, Moscow, Lyublinskaya st., 151, room 222

E-mail: dima_filipchuk@mail.ru

Abstract. The sensitivity of MIS-sensor to the products of thermal decomposition of unsymmetrical dimethylhydrazine was investigated. It is shown that MIS sensor is able to detect the concentrations of the test substance by the means of the certain products of its thermal decomposition (ammonia and nitric dioxide).

1. Introduction

This article is devoted to the investigation of the sensitivity of a MIS-sensor to vapor of unsymmetrical dimethylhydrazine (UDMH) in air. UDMH is one of the components of rocket fuel, which is actively used in Russian space industry. This substance is extremely dangerous for human organism. Even its small concentrations can cause a disruption in the work of internal organs and central nervous system.

There is no simple, affordable and automatic (without human participation) method of detecting the concentrations of UDMH. Available methods (chromatography, ion mobility spectrometry) are very complex and have to use an expensive equipment and high qualification of personnel.

A MIS-sensor can be used to detect concentrations of UDMH for some reasons. It has very high sensitivity to ammonia and nitric dioxide (at the level of parts of ppm [1,2]), which are some of products of thermal decomposition of UDMH [3,4]. If, during the thermal decomposition of UDMH, concentrations of ammonia and nitric dioxide are formed and its values exceed the minimum detectable by a MIS-sensor, then the MIS-sensor can be used to detect the vapor of UDMH in air.

2. Scheme of MIS-sensor

Scheme of MIS-sensor is shown in Figure 1. A silicon plate of KEF-15 (4) coated with a SiO_2 film (3) is used as a basis of the structure. A Ta_2O_5 film (2) is applied to SiO_2 film. Its thickness is about 100 nm. Palladium film (30 nm) is applied to Ta_2O_5 film. The MIS-structure is maintained at a constant temperature of 100-150 °C. Its temperature is measured by thermistor (10) and stabilized by the electronic circuit of the gas analyzer.

A MIS-structure is a capacitor whose capacitance depends on the voltage applied to it (gate voltage). This dependence is called volt-farad (C-U) characteristics. When a certain gas appears above the palladium film, C-U characteristics shifts to the left or right along the voltage axis. It becomes possible to detect the change of MIS-sensor capacitance at a fixed gate voltage (Figure 2) [5,6].

IOP Conf. Series: Journal of Physics: Conf. Series **945** (2017) 012023 doi:10.1088/1742-6596/945/1/012023

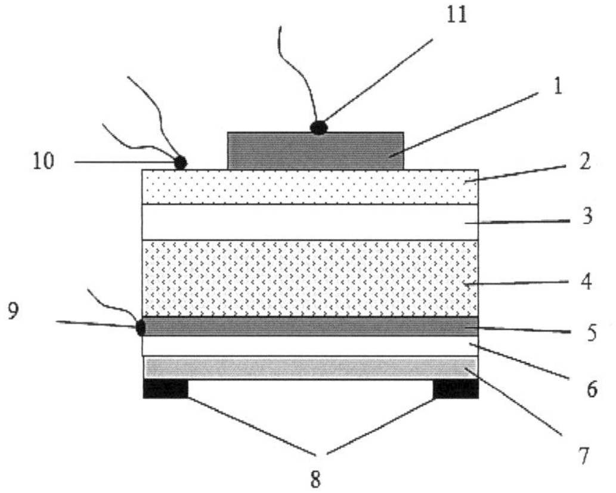

Figure 1. MIS-sensor scheme: 1 – Pd film; 2 – Ta_2O_5 film; 3 – SiO_2 layer; 4 – silicone wafer; 5 – metal electrode; 6 –insulating layer; 7 – heater; 8 – electric contacts of heater; 9,11 – electric contacts of MIS-capacitor; 10 – thermistor.

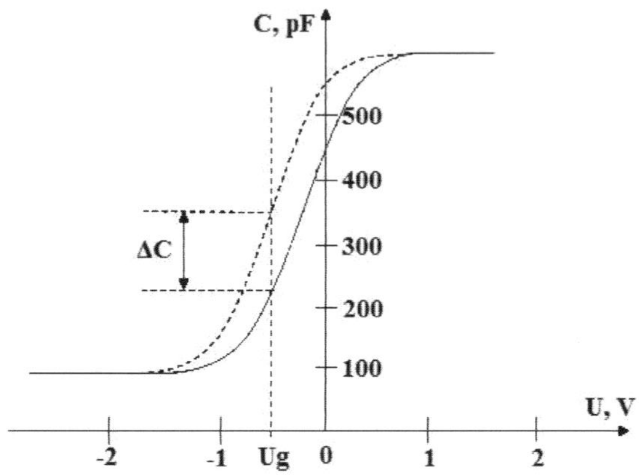

Figure 2. C-U characteristic of MIS-sensor. Dashed line shows the shift of C-U characteristic under the influence of gas.

3. Description of the experiment

An experimental facility was assembled to conduct experiments on thermal decomposition of UDMH. Its scheme is shown in Figure 3.

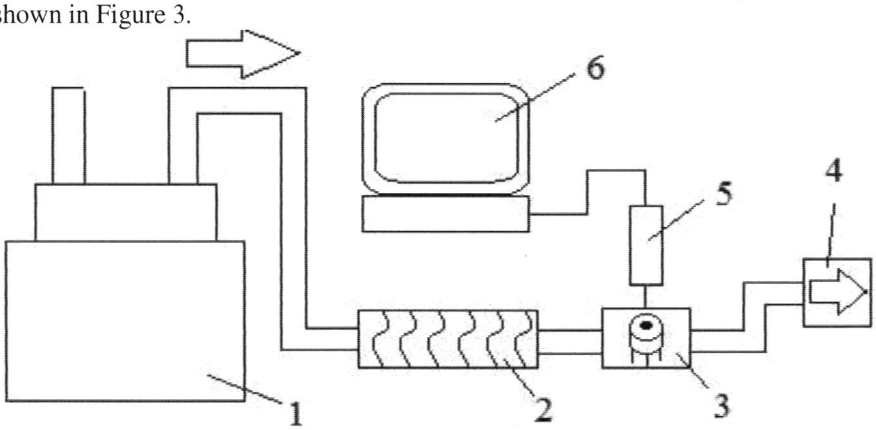

Figure 3. Scheme of facility. 1 – glass container for dilution of UDMH, 2 – reactor, 3 – MIS-sensor, 4 – electric pump, 5 – electronic circuit, 6 – PC.

An experimental facility worked as follows. The concentrations of UDMH was created by dilution in the air of the glass container. Liquid UDMH was used in the work, which was injected into glass container by chromatographic syringe. Liquid UDMH evaporated in the container and then gas mixture of UDMH appeared. By means of the pump the gas mixture was supplied from the glass container to a heated reactor (2). Reactor is a quartz tube wrapped with nichrome wire. When the current passed through the nichrome wire, the air inside the quartz tube was heated to a temperature of 800 $^{\circ}C$ (precisely at this temperature the pyrolysis of UDMH proceeds most efficiently [3]). All products of thermal decomposition were removed by means of airflow to the MIS-sensor (4). The signal from MIS-sensor was processed using the electronic circuit (5) and PC (6).

Before carrying out all experiments, calibration curves of two MIS-sensors for ammonia and nitric dioxide was obtained.

4. Results and its discussion

The ammonia concentration at the outlet of the reactor as a result of the thermal decomposition of UDMH should depend on the concentration of UDMH created in the glass container. To determine the relationship between these values, the following experiment was carried out. In the glass container, concentrations of NDMH 0.015, 0.06 and 0.12 ppm were created, and then the response of the MIS sensor was measured. The values of the sensor response were recalculated into ammonia concentration values using a calibration curve. The results of the experiment are shown in Figure 4.

As can be seen from the graph, as the concentration of UDMH in the glass container increases, the ammonia concentration at the reactor outlet also increases. This confirms the fact of the pyrolysis of UDMH, and also the fact that with the help of a MIS-sensor it is possible to detect concentrations of UDMH by the means of the one of its thermal decomposition products (ammonia).

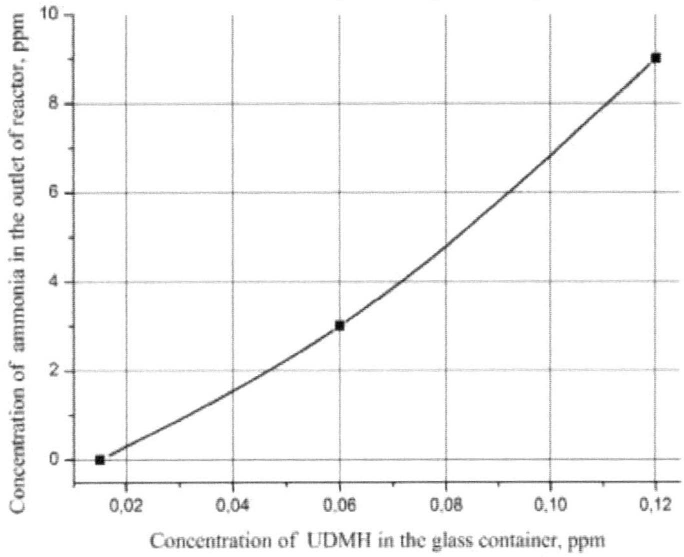

Figure 4. Dependence of ammonia concentration at the outlet from the reactor on the concentration of UDMH, created in a glass container.

The article [4] describes a method for detecting the concentration of UDMH from one of its thermal decomposition products (nitric oxide) at a temperature of 600-1000 °C in the presence of nickel as a catalyst. When nitric oxide is released, it is oxidized to nitrogen dioxide, to which the MIS sensor exhibits a high sensitivity.

Several tubes of nickel foil were installed in the reactor. Then it was heated to 800 °C, and through it a gas mixture was passed from a glass container containing a certain concentration of UDMH (0.06 and 0.12 ppm), and the response of the MIS sensor was measured. Using the calibration curve for nitrogen dioxide, the response values were recalculated to values of nitrogen dioxide concentration. The results of this experiment are shown in Figure 5.

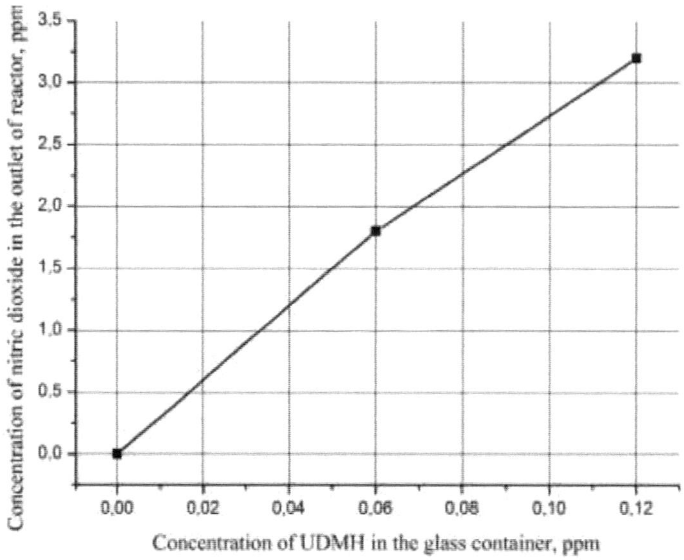

Figure 5. Dependence of nitric dioxide concentration at the outlet from the reactor on the concentration of UDMH, created in a glass container.

As can be seen from the graph, as the concentration of NDMH in the glass container increases, the concentration of nitrogen dioxide at the reactor outlet also increases. This confirms the fact of the pyrolysis of UDMH, and the fact that with the help of a MIS-sensor it is possible to detect concentrations of UDMH by the means of the one of its thermal decomposition products, nitrogen dioxide.

5. Conclusion

Both experiments show that a MIS sensor can be used to detect UDMH vapor by the means of the two products of its thermal decomposition, ammonia and nitrogen dioxide (last one appears only in the presence of nickel), whose concentrations during pyrolysis of UDMH can be easily detected by used MIS-sensors.

References

[1] Emelin E V, Nikolaev I N and Sokolov A V 2005 Sensitivity of MIS-sensors to concentrations of different gases *Sensors and systems* **10** 37-9 (in Russian)

[2] Bolodurin B A, Borisenkov I L, Korchak V Yu, Litvinov A V, Nozdrya D A, Pomazan D V, Filipchuk D V and Etrekova M O 2016 Comprehensive research of sensitivity of MIS-sensors with Pd-SiO2-Si, Pd-Ta2O5-SiO2-Si structures to the content of various gases in air *Russian chemistry journal* **3** 96 (in Russian)

[3] Ushakova V G, Shpigun O N and Starygin O I 2004 Peculiarities of chemical transformation of UDMH and its behavior in environmental objects *Polzunovskij vestnik* **4** 3-6 (in Russian)

[4] Rounbehler D R 1984 Detection of hydrazine compounds in gaseous samples by their conversion to nitric oxide-yielding derivatives US4775633

[5] Nikolaev I N, Litvinov A V and Emelin I N 2006 Model of sensitivity of MIS-sensors to concentrations of different gases *Sensors and systems* **7** 66-73 (in Russian)

[6] Nikolaev I N, Kalinina L I and Litvinov A V 2007 New type of shallow traps for molecules with dipole moments in dielectrics *Solid State physics* **6** 1065-9 (in Russian)

IOP Conf. Series: Journal of Physics: Conf. Series **945** (2017) 012024 doi:10.1088/1742-6596/945/1/012024

Mechanical characteristics of heterogeneous structures obtained by high-temperature brazing of corrosion-resistant steels with rapidly quenched non-boron nickel-based alloys

B Kalin, M Penyaz, A Ivannikov, O Sevryukov, D Bachurina, I Fedotov, A Voennov and E Abramov

National Research Nuclear University MEPhI
Kashirskoe highway 31, Moscow, 115409, Russia

E-mail: MAPenyaz@mephi.ru

Abstract. Recently, the use rapidly quenched boron-containing nickel filler metals for high temperature brazing corrosion resistance steels different classes is perspective. The use of these alloys leads to the formation of a complex heterogeneous structure in the diffusion zone that contains separations of intermediate phases such as silicides and borides. This structure negatively affects the strength characteristics of the joint, especially under dynamic loads and in corrosive environment. The use of non-boron filler metals based on the Ni-Si-Be system is proposed to eliminate this structure in the brazed seam. Widely used austenitic 12Cr18Ni10Ti and ferrite-martensitic 16Cr12MoSiWNiVNb reactor steels were selected for research and brazing was carried out. The mechanical characteristics of brazed joints were determined using uniaxial tensile and impact toughness tests, and fractography was investigated by electron microscopy.

1. Introduction

Recently, the use rapidly quenched boron-containing nickel filler metals for high temperature brazing corrosion resistance steels different classes is perspective. The use of these alloys leads to the formation of a complex heterogeneous structure in the diffusion zone that contains separations of intermediate phases such as silicides and borides [1]. This structure negatively affects the strength characteristics of the joint, especially under dynamic loads and in corrosive environment. Figure 1 shows images in back-scattered electrons (BSE) of brazed joints containing heterogeneities in the diffusion zone.

The work [2] described the development of filler metal based on the Ni-Si-Be system and the method for obtaining a connection by high-temperature brazing.

Nanocrystalline filler metals based on Ni-Cr-Si-Be system were obtained by method of rapidly quenching of a melt [3]. Vacuum high-temperature brazing of austenitic (12Cr18Ni10Ti) and ferritic-martensitic (16Cr12MoSiWNiVNb) steels was carried out with these alloys. The main regularities of the structural-phase state formation of brazed joints and the features of the molten filler metal interaction with brazed materials were revealed, optimum temperature and time parameters of the brazing process were determined.

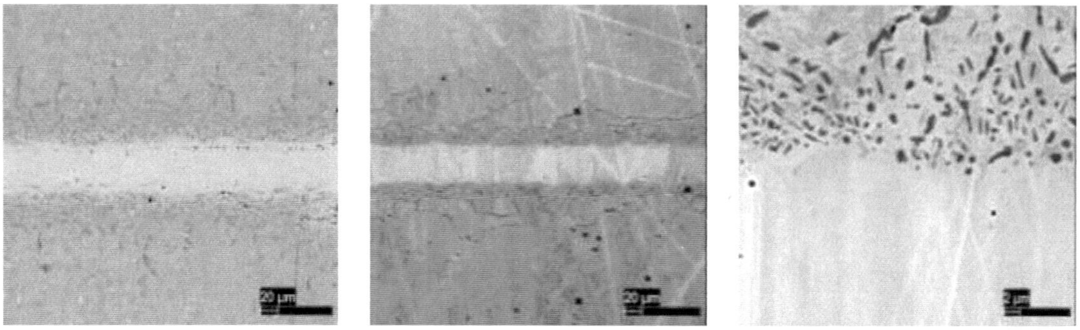

a b c

Figure 1. Brazed joints obtained with the filler metal Ni-7.5Cr-3.5Fe-4.5Si-2.6B wt.%: a) ferritic-martensitic steel 16Cr12MoSiWNiVNb (EP-823), b), c) austenitic steel 12Cr18Ni10Ti (different magnifications).

In the connection obtained with the filler metals containing beryllium, there are no heterogeneities in the diffusion zone because of its significant solubility in steel. It is necessary to compare strength properties to study the operating characteristics of brazed joint obtained with boron and developed non-boron filler metals.

2. Experiment and methods

SSHVE-1.2.5 resistance furnace with tungsten heaters, which allows heating in a vacuum of 10^{-2} Pa, is used for brazing samples.

The method of indentation is used to study the degree of brazed joint heterogeneity. Determination of the microhardness values by Vickers was performed on the installation HVS-1000. The measurement is carried out by pressing a diamond pyramid into the surface of the test sample under the action of a load of 50-100 g, which is applied for 10-15 seconds. Microhardness is measured directly in the seam and at a certain distance from the seam in steps of 50-100 μm. Several measurements are made for each characteristic point.

The strength properties of the samples under uniaxial tension were determined using a universal two-column QUASAR 50 testing machine, which has the maximum force of which is 50 kN. The seams obtained by three filler metals: Ni-7Cr-5Si-3Be, Ni-6Si-5Be and standard boron Ni-7.5Cr-3.5Fe-4.5Si-2.6B (table 1) are tested. They were obtained with the same parameters of cylindrical samples manufacturing in accordance with GOST 28830-90 (working part with a diameter of 6 mm) [4].

Table 1. Brazed specimens for uniaxial tension tests.

Steel	Filler metal	Brazing mode
12Cr18Ni10Ti	Ni-7.5Cr-3.5Fe-4.5Si-2.6B	1150 °C 40 min
12Cr18Ni10Ti	Ni-7Cr-5Si-3Be	1150 °C 40 min
12Cr18Ni10Ti	Ni-6Si-5Be	1150 °C 40 min
16Cr12MoSiWNiVNb	Ni-7.5Cr-3.5Fe-4.5Si-2.6B	1150 °C 15 min
16Cr12MoSiWNiVNb	Ni-7Cr-5Si-3Be	1150 °C 15 min
16Cr12MoSiWNiVNb	Ni-6Si-5Be	1150 °C 15 min

The chemical composition of the using steels is shown in table 2. Brazing of the samples was carried out at the temperature 1150 °C [5]. The holding time for ferritic-martensitic steel is reduced to 15 minutes, compared to austenitic steel to reduce the effect of structural-phase state changes at the brazing temperature that occurs due to the partial dissolution of Nb, V, W and Mo carbides.

The developed filler metal with 50 μm thickness and the boron filler metal with 35 μm thickness were used in one layer and placed directly in the brazing gap. Brazing was carried out in a special conductor.

Table 2. Chemical composition of used steels.

Material	Alloying elements, % wt									
	C	Cr	Si	Mn	Ni	Mo	V	Nb	W	Ti
12Cr18Ni10Ti	0,12	18	<0,8	<2	10	-	-	-	-	<1
16Cr12MoSiWNiVNb	0,15	11,2	1,1	0,7	0,5	0,6	0,2	0,3	0,7	

The fracture zones of the tested samples were analyzed by scanning electron microscopy (SEM) after uniaxial tensile tests. The fractures are investigated in secondary electrons (SE) and back-scattered electrons (BSE) for a more detailed analysis. The profiles of the destroyed samples fractographies after test are also analyzed additionally.

Rectangular samples with dimensions of 10 × 10 × 55 mm without stress concentrators (notches) were made in accordance with the State Standard 23046-78 to determine the impact toughness of brazed joints [6]. The tests were carried out using special apparatus with pendulum. A study of the seam structure before mechanical tests using optical microscopy was conducted to control the quality of the test samples.

3. Results and discussion

The effect of the structural components microhardness on the brazed joints strength [7]. The results are shown in figure 2. The metallographic studies of the brazed joint showed the different structure of the diffusion zone after brazing with boron and beryllium filler metals. The presence of a wide diffusion zone (about 100 μm) with a structure different from steel and seam structure is characteristic for brazed compounds 16Cr12MoSiWNiVNb steel obtained with the Ni-7.5Cr-3.5Fe-4.5Si-2.6B. This zone is formed due to the active diffusion of boron, which leads to recrystallization in steel. The microhardness in this zone is 400±20 HV, which is on 200 HV more than in the joint (figure 2a). The diffusion zone with a size of 100-120 μm is also characteristic for 12Cr18Ni10Ti steel, brazed boron filler metal. However, recrystallization does not occur (figure 2c). Boron forms intermediate phases with chromium along the grain boundaries [8]. Microhardness in this zone falls from 220±10 HV, characteristic of the seam, to 160±10 HV. The difference in the values of microhardness in these regions will lead to the appearance of microstrains.

An eutectic structure is formed in some places in the center of the seam in the case of using the Ni-7Cr-5Si-3Be filler metal. However, its microhardness is close to the microhardness of the nickel matrix in the seam for ferritic-martensitic steel (figure 2b). The microhardness values for austenitic steel from 240±10 HV in the seam fall to 160±10 HV in the diffusion zone and steel (figure 2d).

The structural difference for these two filler metals, as will be seen below, will lead to different results of tests on strength and impact toughness.

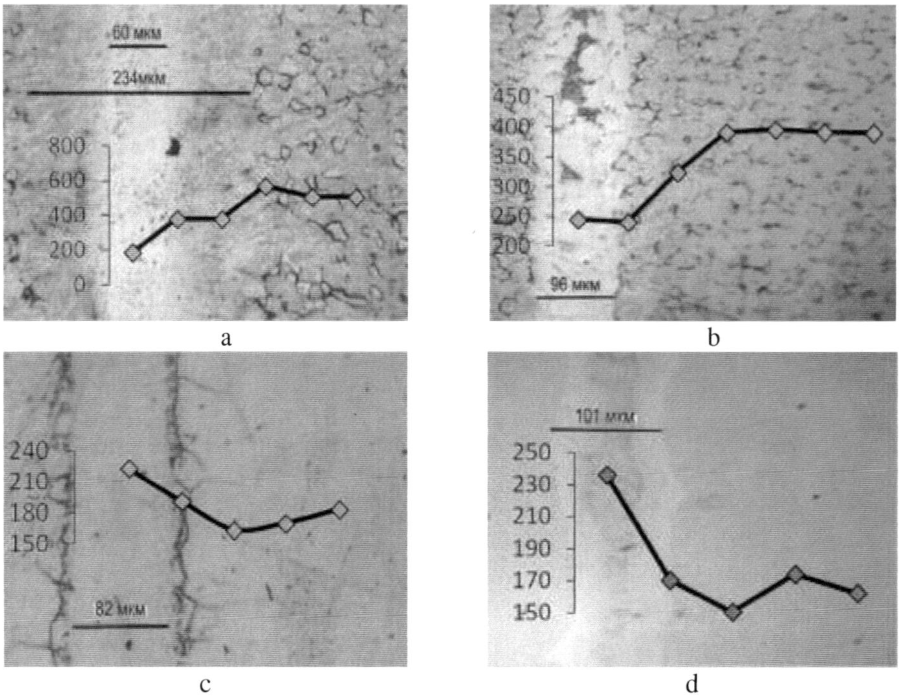

Figure 2. Brazed joint steel 16Cr12MoSiWNiVNb, obtained with filler metal: a) Ni-7Cr-3.5Fe-4.5Si-2.6B; b) Ni-7Cr-5Si-3Be; and steel 12Cr18Ni10Ti: c) Ni-7Cr-3.5Fe-4.5Si-2.6B; d) Ni-7Cr-5Si-3Be.

Figure 3 shows the results of the tests. As a result, recrystallization (in the case of the use of boron filler metal) leads to decrease of the mechanical properties. Therefore, the brazed joint 16Cr12MoSiWNiVNb/Ni-7Cr-3.5Fe-4.5Si-2.6B showed low values of the tensile strength (figure 3a).

The destruction occurred over the steel during the tests of 12Cr18Ni10Ti samples brazed with filler metal Ni-6Si-5Be. Therefore, the joint withstands comparable loads with the base material. The joint obtained with the filler metal Ni-7Cr-5Si-3Be showed the best strength properties. It was also taken into account that steel after processing by the brazing mode changes its properties. It is strengthened and has microhardness higher than in the initial state. This is due to changes in the structural-phase state at the brazing temperature.

The study of the brazed joint fractography showed the following: in the case of boron-containing filler metal, the destruction occurs mainly along the boundary of the seam-base material interface with elements of destruction along the seam body. This indicates its low cohesive strength (figure 4a). Destruction over the body of the seam is predominantly brittle with numerous secondary cracks [9]. In the case of a beryllium-containing filler metal, the destruction of the joint occurs along the seam, which indicates a significantly higher cohesive strength of the seam-base material interface (figure 4b). The nature of the fracture is mainly viscous.

The viscous type of fracture of the sample brazed with beryllium-containing filler metal indicates a high plastic deformation and a significantly higher energy of fracture of this sample as compared to the sample, brazed with boron-containing filler metal and brittle destroyed. The results of fractographic studies are confirmed by the results of the sections investigations (figure 5).

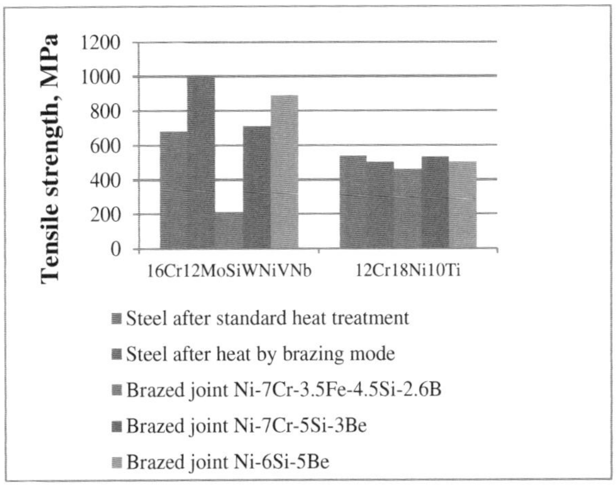

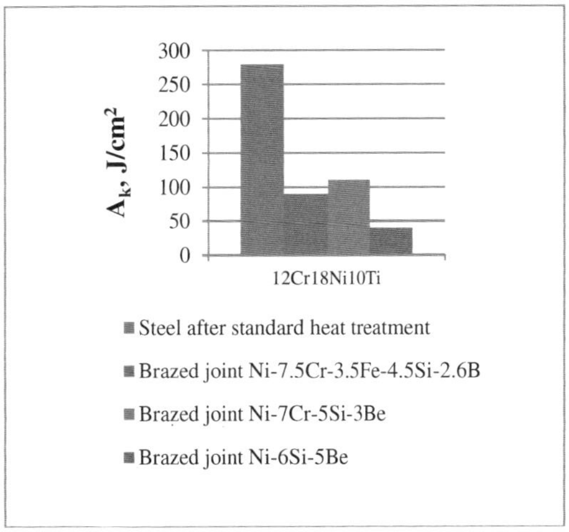

a b

Figure 3. The results of tests on a) uniaxial tensile; б) impact toughness for steel 12Cr18Ni10Ti.

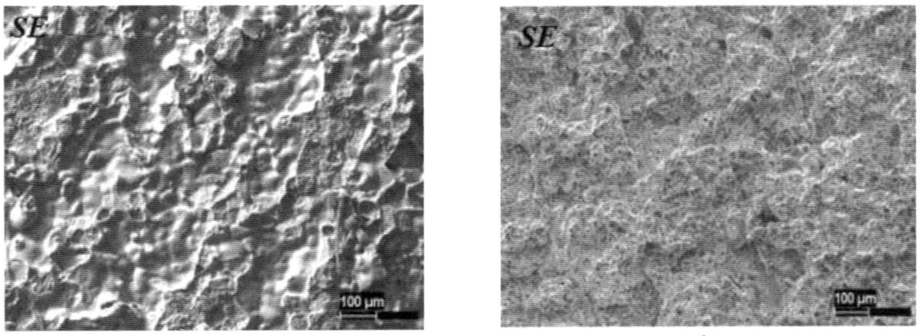

a b

Figure 4. Fractography (SE) of destroyed joint steel 16Cr12MoSiWNiVNb, obtained with the filler metal: a) Ni-7Cr-3.5Fe-4.5Si-2.6B; b) Ni-6Si-5Be.

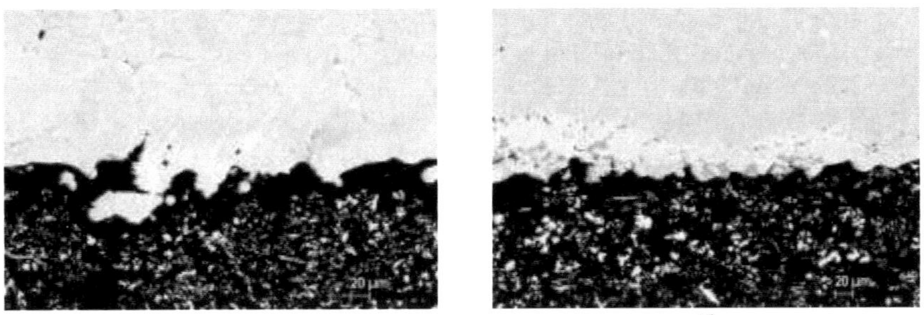

a б

Figure 5. Microstructures of destroyed joint steel 16Cr12MoSiWNiVNb, obtained with the filler metal:-a) Ni-7Cr-3.5Fe-4.5Si-2.6B; b) Ni-6Si-5Be.

It can be seen from the figure that the grain is growing strongly due to recrystallization during brazing with boron filler metal. Moreover, intermediate phases $(Cr,Fe)_2B$ are formed along the boundaries, which weaken these boundaries, therefore, destruction occurs subsequently on them. Diffusion of beryllium, unlike boron, does not lead to recrystallization in steel. It, significantly dissolving, is found both along the boundaries and in the body of grain.

Unlike the results of uniaxial tensile tests, the results on the impact toughness obtained for joint of 12Cr18Ni10Ti steel brazed with Ni-7Cr-5Si-3Be filler metal are higher than for Ni-6Si-5Be and Ni-

7Cr-3.5Fe -4.5Si-2.6B. The worst results were shown by the filler metal Ni-6Si-5Be because of the large content of beryllium, which reduces the ductility of the joint (figure 3b).

It can be concluded that the optimal composition for the developed alloy is Ni-7Cr-5Si-3Be, which demonstrates high values of strength and plasticity.

4. Conclusions

High-temperature brazing of austenitic (12Cr18Ni10Ti) and ferritic-martensitic (16Cr12MoSiWNiVNb) steels with classical boron filler metal Ni-7Cr-3.5Fe-4.5Si-2.6B and developed filler metal based on the system Ni-(Cr)-Si-Be was conducted. Joints were tested for uniaxial tensile strength and impact toughness. Analyzing the obtained data, it can be concluded that the developed filler metal based on the Ni-(Cr)-Si-Be system demonstrate high strength characteristics of brazed joints. The most promising was the composition Ni-7Cr-5Si-3Be. The characteristics obtained for it: tensile strength for 12Cr18Ni10Ti 530±30 MPa, for 16Cr12MoSiWNiVNb 710±30 MPa; impact toughness for 12Cr18Ni10Ti 110±10 J/cm^2. The problem of the negative influence of the heterogeneous structure in the near-seam zone (the "boride grid") is solved by replacing the boron with beryllium in the composition of the filler metal. However, a large content of beryllium in the alloy (as in the case of Ni-6Si-5Be) is also undesirable, since the plasticity of the joint decreases. The microhardness of the compound was investigated to determine the influence of the structural components of the joint on the strength properties. The possibility of using the developed filler metal Ni-7Cr-5Si-3Be for connecting energy-stressed elements from steels of various classes is shown.

References

[1] Maksimova S V 2007 Amorphous filler metals for brazing stainless steel and titanium and the structure of brazed joints *Adhesion of melts and brazing of materials* pp 70-81

[2] Ivannikov A and Kalin B 2016 Development of rapidly quenched nickel-based non-boron filler metals for brazing corrosion resistant steels *IOP Conference Series: Materials Science and Engineering* **130** (2016) 012044

[3] Ivannikov A and Fedotov V 2016 The use of Ni-Cr-Si-Be filler metals for brazing of stainless steels *IOP Conference Series: Materials Science and Engineering* **130** (2016) 012045

[4] GOST (State Standart) 28830-90 Brazed joint. Test methods for tensile strength and long-term strength

[5] Ivannikov A and Tukhbatov A 2014 Rapid-quenched nickel-based solder for high-temperature brazing of various constructive elements *Tsvetnye Metally* **12** 27-31

[6] GOST (State Standart) 23046-78 Brazed joint. Method of impact tests

[7] Binesh B and Jazayeri Gharehbagh A 2016 Transient Liquid Phase Bonding of IN738LC/MBF-15/IN738LC: Solidification Behavior and Mechanical Properties *Journal of Materials Science and Technology* **32** 1137-51

[8] Bai K and Ng F 2017 Understanding non-parabolic solidification kinetics in Ni-based alloys during TLP bonding via thermo-kinetic modelling *Journal of Alloys and Compounds* **699** 1084-94

[9] Fellows J 1982 *Fractografy and Atlas of Fracture Patterns: Reference Book* (Moscow: Metallurgiya)

IOP Conf. Series: Journal of Physics: Conf. Series **945** (2017) 012025 doi:10.1088/1742-6596/945/1/012025

Thermal contact resistance of a liquid lead and structural steel

A B Kruglov, V B Kruglov, V S Kharitonov, P G Struchalin and D P Shornikov

National Research Nuclear University "Moscow Engineering Physics Institute", Moscow, Russia

E-mail: PGStruchalin@mephi.ru

Abstract. The paper presents numerical investigation of the thermal resistance of the contact surface of heat exchange and melt of lead. Using a geometric model of the contact area, developed on the basis of image analysis of the contact surfaces of steel and lead, the calculation of the contact resistance between molten lead and the surface of steel EP-823. The results of the calculations are compared with experimental data on thermal resistance of contact of the lead with steels EP-823 and 312 steel.

1. Introduction

The future of nuclear energy is associated with the development of reactor installations on fast neutrons with sodium (BN) and lead (BREST) coolants. The use of such installations nitride nuclear fuels will improve the fuel efficiency and move to a closed fuel cycle.

The fast reactors are characterized by high heat flows and temperatures in the core reactor. At temperatures above 1000 °C nitride fuel begins to rapidly swell due to the increase of fission gas release. Achieving deep burnup of the fuel in such conditions is possible when you use fusible heat-conducting layers between the fuel and sheath, in particular, layers of lead and its alloys. The use of thermally conductive layers reduces the temperature of the fuel, it's the swelling and gas release increases the radiation resistance of fuel rods. High chemical activity of lead, leading to significant corrosion of its interaction with the shell, can be reduced by introducing some additives lead [1].

The calculations showed that in the fuel elements, when the gap between fuel and shell of 0.1 mm, the thermal resistance of liquid metal thermal interface layer is comparable to the thermal resistance of the boundaries of contact of the melt with fuel and shell. With increasing burnup, the thermal conductivity of the fuel will decrease and the thermal resistance will increase in connection with the accumulation on the surfaces of the fuel and shell of gaseous fission products and corrosion products of the interaction of the liquid metal, the shell and fuel. As a result, the temperature in the center of the fuel will increase.

The above factors should be taken into account in the calculations of the thermal regime of the fuel rods and the reactor installation as a whole, and the thermal resistance of the contact boundary layer of liquid metal fuel and the cladding shall be experimentally investigated.

2. The content of this work

Earlier, in [2, 3] conducted the experimental study of thermal contact resistance of molten lead, and steel EP-823 (figure 1).

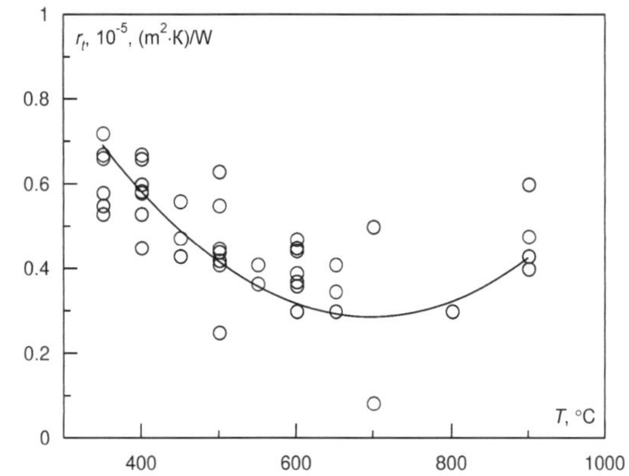

Figure 1. Thermal resistance of the boundaries of contact "steel – liquid lead"

The data shows that the thermal contact resistance r_t " liquid lead - steel" is comparable to the thermal resistance of the layer of lead with a thermal conductivity λ and thickness δ: $r_{Pb} = \delta/\lambda$. When $\lambda = 20$ W/(m·K) and $\delta = 0.1$ mm (this value of the gap assumed in the fuel rods with nitride fuel) we obtain the value of $r_{Pb} \approx 0.5{\cdot}10^{-5}\,(m^2{\cdot}K)/W$.

In the present work for the analysis of the measurement results of thermal resistance, a study was conducted of the structure of surfaces of lead and steel in the contact area with electronic and optical microscope. The contact area was formed in an argon atmosphere on a steel substrate during melting at $T = 350$ °C, soaking at a temperature of 650 – 900 °C for 0.5 – 1.5 hours and cooling to room temperature. Then a drop of lead was separated from the substrate and carried out the study of the contact surfaces of lead and steel, optical and electronic (Jeol 6610 LV) microscopes. Figures 2 and 3 show examples of the obtained images. Analysis of the images revealed many areas of local wetting of the surface of steel by lead. The proportion of the wetted surface is small and the greater part of the surface of contact with lead, such as lead was separated from the surface by layer of gas.

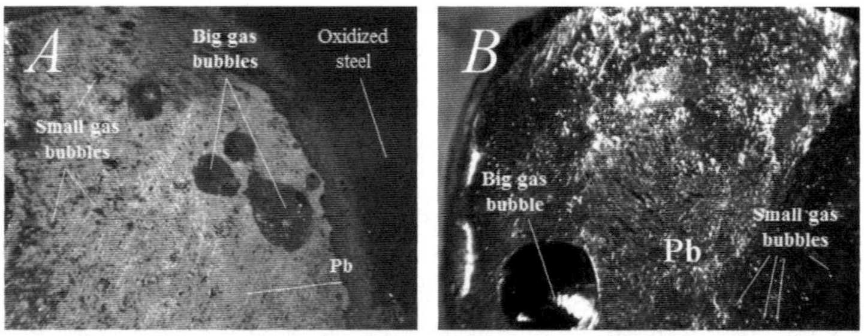

Figure 2. Contact surfaces of steel (*a*) and lead (*b*) (Pictures on optical microscope)

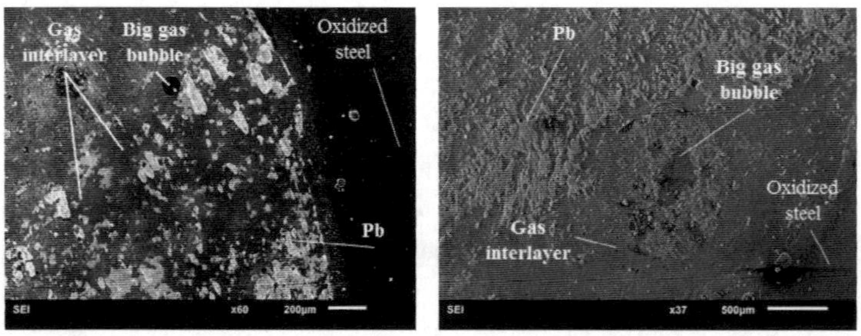

Figure 3. Contact boundary of steel (Pictures on electronic microscope)

On a substrate made of steel EP-823 by the method of lying drop studied the change in wetting angle by liquid lead at the temperature of crystallization of $T_k = 327{,}5$ C depending on temperature and exposure time of the melt on the steel surface. In argon atmosphere was studied wetting of the two surfaces: brushed (No. 1); polished, then oxidized (No. 2). By results of measurements it is established that the wetting angle $\alpha_k(T_k)$ steel EP-823 by lead decreases with increasing temperature and the total exposure time and stabilized on level $\alpha_k(T_k) \approx 100°$ (figure 4).

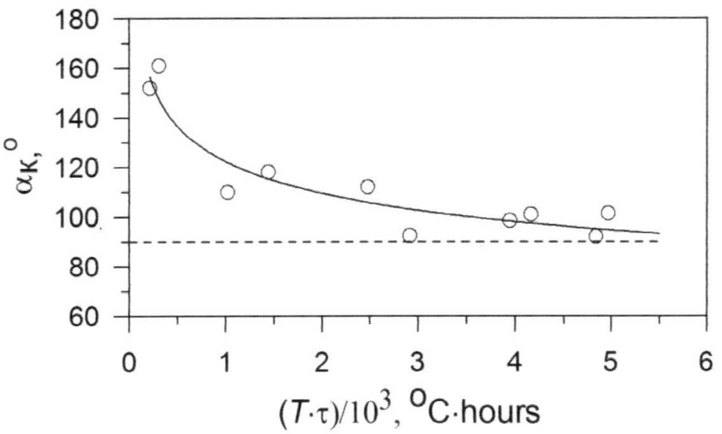

Figure 4. The dependence of the wetting angle became the lead from the time and temperature of exposure.

After analyzing the images of the contact surfaces we created the model of contact of the molten lead and the steel surface (figure 5). According to this model, the contact consists of areas of the snug lead to the steel through the oxide film and areas with a gas layer between the liquid lead and steel. In the model, the heat flux transmitted through the boundary between steel and molten lead, encounters the thermal resistance of a single oxide layer, or together with the resistance of the gas layer. Thermal resistance of the oxide layer in its thickness $\delta_{ox} \sim 1$ mkm and $\lambda_{ox} \sim 1$ W/(m·K) estimated to be $r_{ox} = \delta_{ox}/\lambda_{ox} \sim 10^{-6}$ (m²·K)/W. The thermal resistance of the gas layer with thickness comparable to the mean free path of the gas molecules, determined by the accommodation of gas molecules to the surfaces of lead and steel [4]. In the conditions of the experiments the calculation of the thermal resistance of the gas layer leads to the values $r_g \sim 10^{-4}$ (m²·K)/W. Since the thermal resistance of the gas layer is several orders of magnitude more thermal resistance of the oxide layer, the latter can be neglected in places of contact of lead and steel through the gas layer. In this case, the total thermal resistance of a boundary of contact of steel and lead can be recorded with the help of the interpolating formula (1):

$$r_t = \frac{r_{ox} r_g}{(1-\varepsilon_g) r_g + \varepsilon_g r_{ox}} \qquad (1)$$

where r_{ox} – thermal resistance of oxide layer with thickness δ_{ox} and thermal conductivity λ_{ox}, r_g – thermal resistance of gas layer, ε_g – the percentage of the surface covered with a gas layer.

The results of the calculation of the thermal resistance r_t of the proportion of the surface covered with gas ε_g, shown in figure 6.

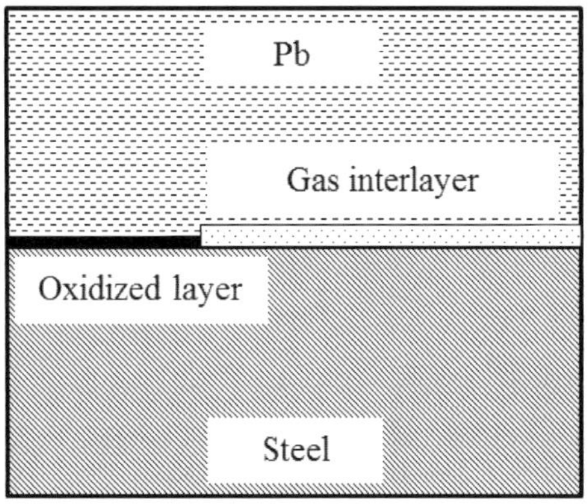

Figure 5. Contact model of a melt of lead and steel

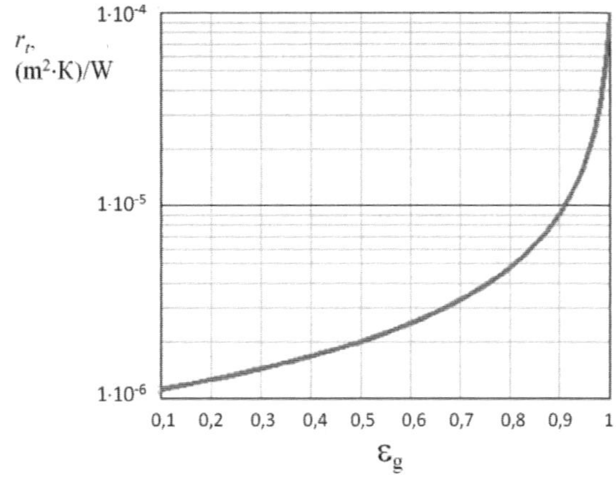

Figure 6. Calculation of thermal resistance r_t

From figure 5 it is seen that the thermal resistance $r_t \approx 0.5{\cdot}10^{-5}$ (m^2·K)/W, observed in experiments, corresponds to the fraction of the gas interlayer $\varepsilon_g \approx 0.8$. Analysis of the obtained images of the surfaces of contact of steel and lead is consistent with this value ε_g.

3. Conclusion

The study of the structure of the contact liquid lead and structural steel developed a model of contact geometry on the basis of which the calculations of the thermal resistance of the contact. The obtained results agree with experimental data.

The developed model can be used to predict changes in thermal resistance due to the accumulation of gaseous fission products in the thermal interface layer between the fuel and shell in the fuel element during operation in the reactor.

4. Acknowledgments

This work was supported by the Russian Science Foundation, project no. 16-19-10548.

References

[1] Orlova E A, Kruglov A B, Chuvaev D V, Struchalin P G, Zagrebayev S A, Zhmurin V G, 2016 *VANT. Series: nuclear reactor constants* **4** 95–8 (In Russian)

[2] Kruglov A B, Kruglov V B, Struchalin P G, Kharitonov V S, Orlova E A, Zagrebayev S A, Zhmurin V G, 2016 *Bulletin of the Lebedev Physics Institute* **43** 302-5

[3] Kruglov A B, Kruglov V B, Struchalin P G, Kharitonov V S, 2015 *Abstracts of reports at scientific conference* (Obninsk: SSC RF-IPPE) 274 (In Russian)

[4] Kirillov P L, Bobkov V P, Zhukov A V, Yuryev Yu S *Reference book on thermohydraulic calculations in nuclear power* (vol 1) ed P Kirillov (Moscow, Izdat) p 776 (In Russian)

IOP Conf. Series: Journal of Physics: Conf. Series **945** (2017) 012026 doi:10.1088/1742-6596/945/1/012026

Solar flare activity in 2006 - 2016 according to PAMELA and ARINA spectrometers

S A Rodenko*, I K Borkut, A G Mayorov, V V Malakhov on behalf of PAMELA collaboration

National Research Nuclear University MEPhI (Moscow Engineering Physics Institute), Kashirskoe highway 31, Moscow, 115409, Russia

*E-mail: sarodenko@mephi.ru

Abstract. From 2006 to 2016 years on the board of RESURS-DK1 satellite PAMELA and ARINA cosmic rays experiments was carried out. The main goal of experiments is measurement of galactic component of cosmic rays; it also registers solar particles accelerated in powerful explosive processes on the sun (solar flares) in wide energy range. The article includes the list of solar events when PAMELA or ARINA spectrometers have registered increasing of proton flux intensities for energies more than 4 MeV.

1. Introduction

Solar cosmic rays (SCR) are forming on the sun by accelerating solar wind plasma particles during solar flares, coronal mass ejections and other energy releasing events. After that they can reach the Earth. Despite a long history of SCR measurements (starting from 1942) there is still a lack of experimental data, especially in relativistic energies ($T > 1$GeV), where just 71 such events was registered in the recorded history [1].

At the same time accurate measurements of cosmic rays spectra in the wide energy range is necessary for study principles of particles generation and acceleration on the sun, their propagation through the interplanetary space, penetration into the Earth magnetosphere and filling radiation belts.

Low energies SCR could be measured just by simple probes intended for radiation situation and space weather monitoring [2, 3], and high energies SCR indirectly with ground stations [4]. PAMELA and ARINA experimental data fill SCR energy range from dozens of MeV to dozen of GeV [5, 6], which hasn't been covered before.

Most important task for cosmic research is providing radiation safety for a spaceship crew and electronics during near space orbiting and interplanetary flights. Solar cosmic rays are the most hazardous and the least predictable component of space weather. Their flux can increase by orders of magnitude in several minutes and keep high during from several hours up to several days.

This is the reason to study SCR influence to the Earth radiation belts. For example in the article [7] (the article is about the Forbush decrease associated with solar flare 17 May 2012) the Earth magnetic field disturbance due to the solar flare is studied. An additional filling with protons and electrons of radiation belts can take place during solar flares. β-decay of free neutrons produced in the interaction of high energy SCR with the Earth atmosphere or violation in adiabatic invariants due to magnetic field disturbance resulting in acceleration of trapped particles and their transit from outer to inner radiation belt.

2. PAMELA and ARINA spectrometers

The satellite RESURS-DK1 with PAMELA and ARINA spectrometers on the board was launched on 15 June 2006 year. Its orbit was semipolar with inclination about 70 degrees, elliptical with apogee 600 km and perigee 350 km until September 2010 when it was changed to almost circular with radius about 600 km.

The PAMELA instrument consist of set of subdetectors: Anticoincidence system is used for exclude particles outside instrument acceptance; Time of Flight system is used for providing trigger signal and to measure velocity; silicon microstrip tracking system surrounded by permanent magnet is used for measure rigidity of particles by deflection in magnetic field; silicon-tungsten calorimeter is used for hadron/lepton separation by shape of shower; neutron detector and shower tail catcher scintillator S4 is to used enhance rejection power for high energies. ARINA spectrometer is simpler device consisted of 10 scintillator plates and is able to register protons end electrons, measuring their energy by measure a number of plates it passes through. Protons energy range available for ARINA is from 45 MeV to 100 MeV. Energy resolution of ARINA and PAMELA spectrometers is about 15% and several percents respectively. A detailed description of the apparatus can be found in [5, 6].

3. Numbers of registered events SCR

Numbers of solar events registered by PAMELA and ARINA spectrometers are in the tables 1 and 2 respectively divided by minimum energy of particles and their type. In the tables one can see that there is a large statistics of particle events to analyze accumulated by spectrometers including four events with protons more than 1 GeV (GLE 70 and 71) and five events with helium nuclei.

Table 1. SCR event numbers: PAMELA

Proton channels, flares		Helium channels, flares	
> 200 MeV	31	> 100 MeV/n	4
> 500 MeV	18	> 500 MeV/n	1
> 1 GeV	4		

Table 2. SCR events numbers: ARINA

Proton channels, flares	
> 45 MeV	58
> 100 MeV	58

Table 3. SCR proton event statistics for different years

	ARINA			PAMELA	
	> 45 MeV	> 100 MeV	> 200 MeV	> 500 MeV	> 1000 MeV
2006	4	4	2	2	2
2011	11	10	5	2	0
2012	27	18	13	7	1
2013	22	14	6	3	0
2014	20	10	4	3	1
2015	11	2	1	1	0

The number of solar events registered by the PAMELA and the ARINA apparatuses distributed by years is in the table 3. Two last strong solar flares of 23rd cycle including GLE 70 was registered by the instruments in 2006 year. No solar events were registered in the further years till March 2011. The time level of solar activity was unusually low. The highest number of solar events was observed in 2012 including GLE 71. High Energy particle solar events also were observed in January 2014. There was no strong solar flare in 2015 and 2016.

Achieved solar event data allows performing physical analysis of as single flares [7, 8] as their sets.

IOP Conf. Series: Journal of Physics: Conf. Series **945** (2017) 012026 doi:10.1088/1742-6596/945/1/012026

4. High energy protons during the solar flare on 23th of January in 2012.

In the figures 1-3 at the top panels one can see differential energy spectra of protons accelerated during solar flare with class M8.7 on 23th January 2012 year in the different stage of particles coming. The moments of time when measurements of spectra took place are marked by red dotted line in the bottom panels. In the top panels intensities of proton fluxes measured by PAMELA and ARINA spectrometers are shown with red squares and light blue circles correspondingly. Galactic cosmic rays spectrum is marked with black squares (measured by PAMELA) and white circles (measured by ARINA).

In the bottom panel time profiles (from 22 Jan to 31 Jan 2012) of proton flux intensities if different energies are shown (PAMELA data marked as black squares, ARINA data as white diamond). Green marks correspond to 45 ÷ 55 MeV proton flux intensity; black marks – 90 ÷ 115 MeV, red marks – 620 ÷ 670 MeV.

Results of PAMELA and ARINA instruments are agreed at each separate moment of time, what validate both measurements, performing independent tests of each, and showing opportunity of study of particle arriving dynamics into the near Earth space in the wide energy range. We also want to note that features of the satellite orbit allow studying cosmic rays particle penetration into the Earth magnetosphere.

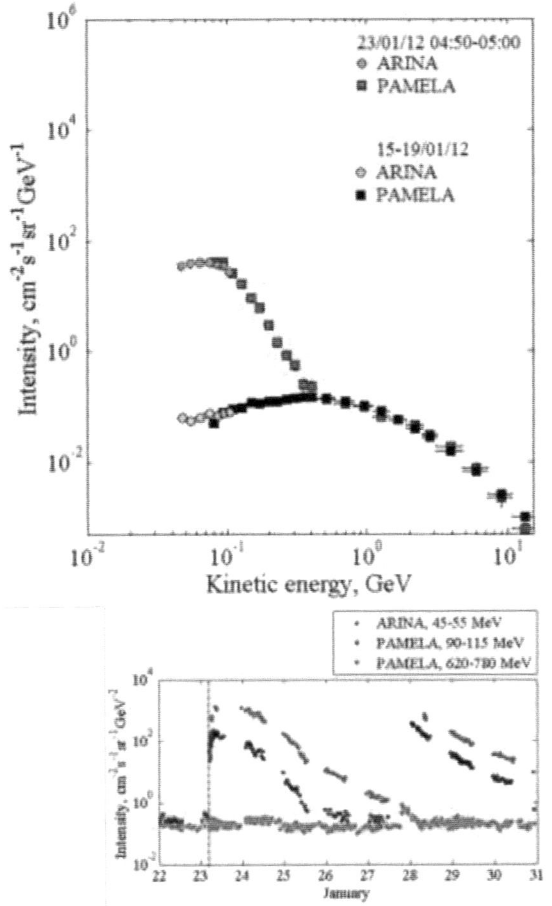

Figure 1. SCR flux obtained on 23.01.12 (04:50 ÷ 05:00)

5. Conclusion

PAMELA and ARINA experiments expand energy range of direct measurements of SCR from dozen of MeV up to the order of GeVs. During the period from June 2006 to January 2016 spectrometers have registered a large number of SCR events reached the Earth's orbit. We plane to conduct their analysis in the future.

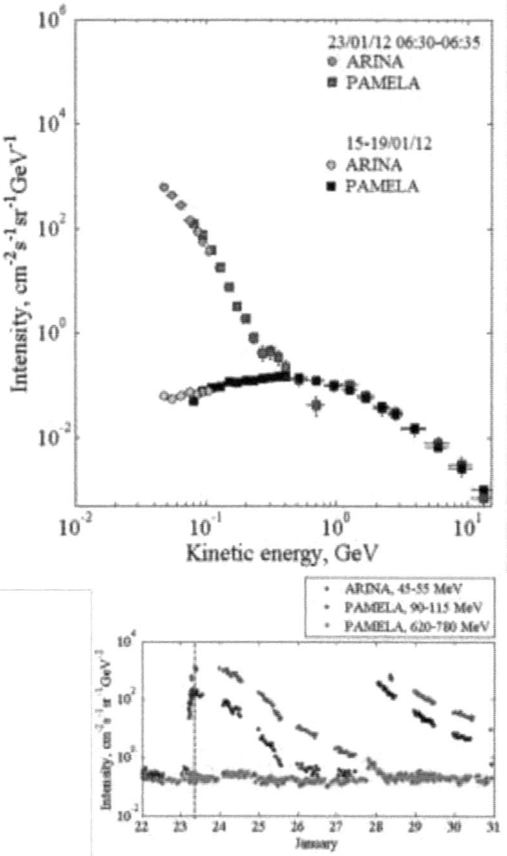

Figure 2. SCR flux obtained on 23.01.12 (06:30 ÷ 06:35)

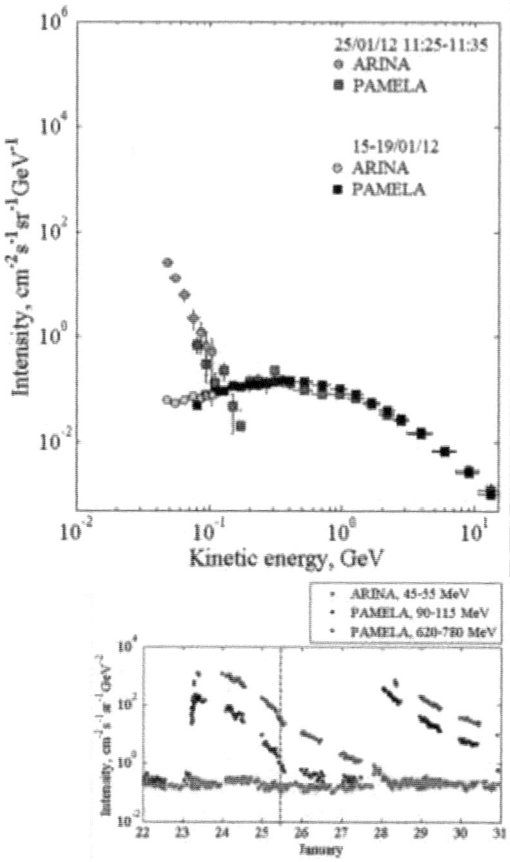

Figure 3. SCR flux obtained on 25.01.12 (11:25 ÷ 11:35)

Acknowledgments
This work was supported by Russian Ministry of Education and Science in frame of project № 3.2131.2017 and by Russian Scientific Foundation (Grant 17-72-20086).

References

[1]. http://cosmicrays.oulu.fi/GLE.html
[2]. http://www.srl.caltech.edu/ACE/
[3]. https://www.ngdc.noaa.gov/stp/satellite/goes/
[4]. https://cosmicrays.oulu.fi/
[5]. Picozza P *et al.* 2007 *Astropart. Phys.* **27** 296–315
[6]. Batishchev A G *et al.* 2005 *Bull. Russ. Acad. Sci.: Physics* **69**(6) 1021–34
[7]. Bruno A *et al.* 2016 *J. Phys.: Conf. Ser.* **675** 032006
[8]. Adriani O *et al.* 2011 *ApJ* **742**(2) 102

Methods of reconstruction of multi-particle events in the new coordinate-tracking setup

V S Vorobyev, V V Shutenko, E A Zadeba

National Research Nuclear University MEPhI (Moscow Engineering Physics Institute), Moscow 115409, Russia

VSVorobev@mephi.ru

Abstract. At the Unique Scientific Facility NEVOD (MEPhI), a large coordinate-tracking detector based on drift chambers for investigations of muon bundles generated by ultrahigh energy primary cosmic rays is being developed. One of the main characteristics of the bundle is muon multiplicity. Three methods of reconstruction of multiple events were investigated: the sequential search method, method of finding the straight line and method of histograms. The last method determines the number of tracks with the same zenith angle in the event. It is most suitable for the determination of muon multiplicity: because of a large distance to the point of generation of muons, their trajectories are quasiparallel. The paper presents results of application of three reconstruction methods to data from the experiment, and also first results of the detector operation.

1. Introduction

The large coordinate-tracking detector based on drift chambers (DC) [1] for investigations of muon bundles generated by ultrahigh energy primary cosmic rays is developed at the Unique Scientific Facility NEVOD. The purpose of the installation is solving the problem of the excess of muon bundles, the magnitude of which increases with the energy of primary cosmic rays that can be caused both by cosmophysical and nuclear physical reasons [2]. The only characteristic of cosmic rays, which reacts differently to changes in the composition of cosmic rays and to the inclusion of new physical processes, but which has not been investigated to date, is the energy of the muon component of EAS. Existing coordinate detector DECOR together with Cherenkov water detector (CWD) NEVOD [3] measures multiplicity and energy release of muon bundles. The supermodules of DÉCOR do not cover the full side of the CWD so some particles can pass between them, the two track resolution of DECOR is 3 cm, these complicate muon bundle study. The new detector will cover entire side of the water volume of the CWD with the possibility of distinguishing close tracks at a distance of 4-5 mm.

At the moment, the coordinate-tracking unit based on drift chambers (CTUDC) has been developed [4]. It is a prototype of the full-scale installation. Detector contains 16 drift chambers forming 2 coordinate planes by 8 DC at opposite sides of the CWD (figure 1). The total area of the detector is 29.6 m^2.

Multiwire drift chambers of CTUDC have been developed in Institute of High Energy Physics (IHEP) [5]. The chamber has an array of field-forming wires arranged on the perimeter, 10 cathode and 4 signal wires in the center (figure 2). The signal wires are alternately shifted

IOP Conf. Series: Journal of Physics: Conf. Series **945** (2017) 012027 doi:10.1088/1742-6596/945/1/012027

from the center by 0.75 mm to solve left-right ambiguity. Drift chamber allows reconstructing the track on a plane perpendicular to the signal wires. The measured drift times of electrons to each signal wire let to determine the ionization points along the particle track, uniformity of electric field allows considering the drift velocity constant.

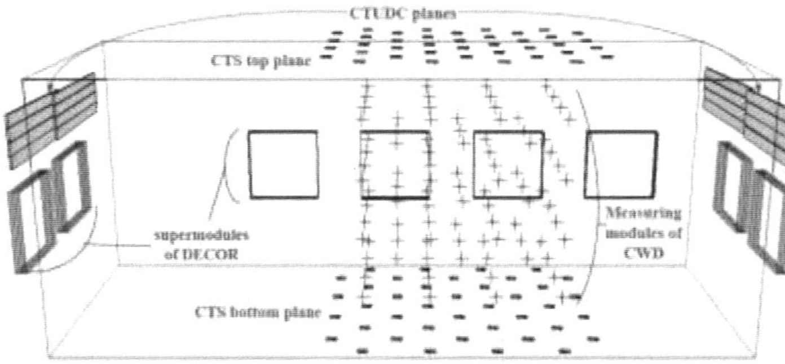

Figure 1. Location of planes of CTUDC relative to CWD, DECOR and calibration telescope system (CTS).

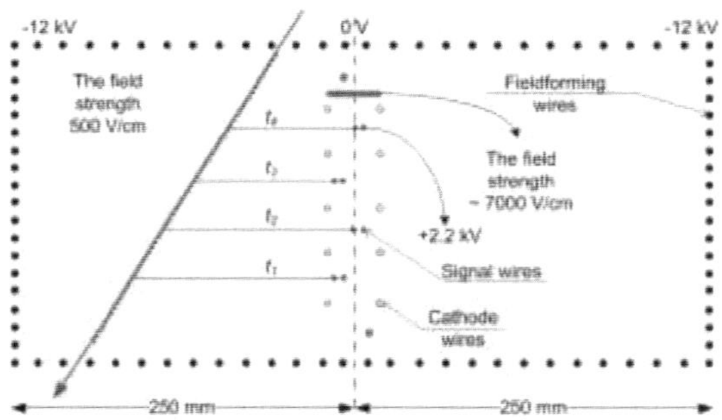

Figure 2. The principle of detecting a track of a particle in a multiwire IHEP drift chamber.

2. Methods of reconstruction

Three methods were applied for reconstruction of multi-particle events registered by drift chamber: the sequential search method, method of finding the straight line and method of histograms.

The sequential search method enables to reconstruct each possible combination of the points using the least square method (LSM) and choose the combination with minimal sum of the squared deviations. After finding the first best track, the next one is selected from the remaining points until at least one channel has no unused points.

Method of finding of the straight line draws straight lines between each pair of points on two selected signal channels. Then each line is extrapolated to remaining channels. If the ionization points are within the allowed region ΔX around the line, reconstruction using LSM is carried out.

Method of histograms was developed for finding and reconstruction of parallel tracks. It draws corridors with a width ΔX at an angle α through each point on the first channel to the others (figure 3). If there is at least one point on each channel in the corridor, all this points are reconstructed by a sequential search method. All α angles are enumerated in the range from 25° to 155°, for each value the number of found tracks is counted. The chosen reconstruction of an event is the one for which the greatest number of tracks is found.

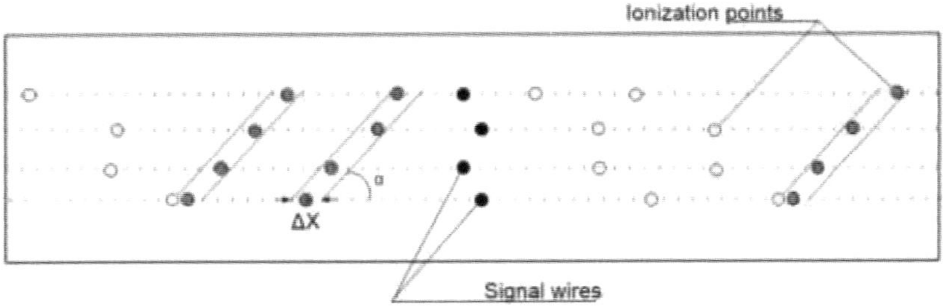

Figure 3. Reconstruction by the method of a histogram.

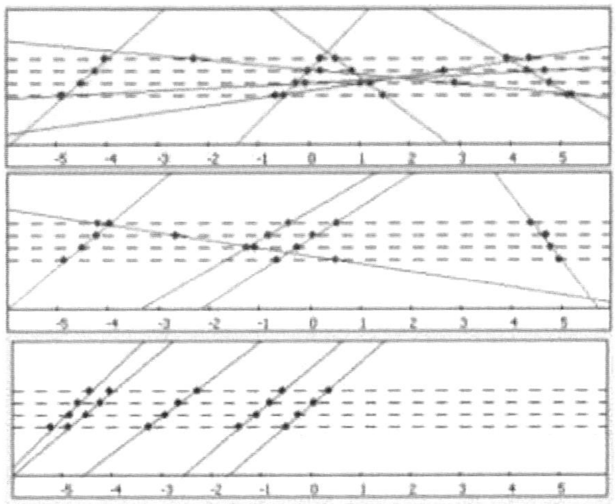

Figure 4. Example of the event reconstruction by different methods: red – the sequential search method, green – method of finding the straight line, blue – method of a histogram.

The Figure 4 shows comparison of reconstructions of an event containing a bundle of parallel tracks by three methods. Reconstruction with the first method is well suited for events with small multiplicities, analysis of drift chamber characteristics for single-particle events, and calibration of other methods. For events with a large multiplicity, the second and third methods are used. Since the muon bundles are quasi-parallel, preference is given to the histogram method. This allows us not only to find the maximum number of tracks of muons, but also to partially reject tracks of secondary particles.

3. Reconstruction of parallel muon bundles in CTUDC

Reconstruction with method of histograms in CTUDC and parallel tracks counting are carried out simultaneously in all chambers. The selection of events is carried out according to the following criteria: the minimum number of tracks is 3, tracks are reconstructed in at least two chambers.

Also, a comparison of joint events registered with the CTUDC and DECOR installations was conducted. The Figure 5 shows the reconstruction of one event in these installations. As seen from the figure, the reconstructions fit each other. At this stage, the angles in the joint events coincide with an accuracy of about 4°. Further improvements of the method and chamber calibration will allow to reduce this difference.

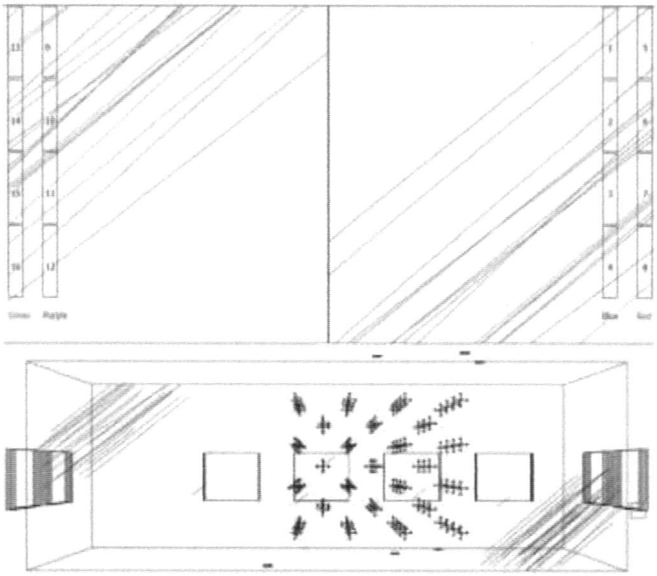

Figure 5. Comparison of the reconstruction of the event in CTUDC (up) and DECOR (down). The difference in the definition of the angle is 1°.

4. Conclusion

For the reconstruction of multi-particle events in the IHEP drift chambers, the sequential search method, method of finding the straight line and method of histogram were applied. The main purpose of the new coordinate-tracking setup is to determine the zenith angle and multiplicity of the muon bundles, for this purpose the method of histograms has been found the most suitable.

The method was studied using simulation and experimental data. The comparison of reconstructions of multiple events based on the data of CTUDC and DECOR showed the correctness of the method.

Acknowledgments

This work was performed on the Unique Scientific Facility NEVOD. It was supported by the RF Ministry of Education and Science (project RFMEFI59114X0002 and MEPhI academic excellence project 02.a03.21.0005).

References

[1] Zadeba E A et al. 2014 The coordinate-tracking detector based on the drift chambers for ultrahigh-energy cosmic ray investigations *J. Instrum.* **9** C08018

[2] Bogdanov AG et al. 2010 Investigation of the properties of the flux and interaction of ultrahigh-energy cosmic rays by the method of local-muon-density spectra *Physics of Atomic Nuclei* **73** 852.

[3] Petrukhin A A 2012 NEVOD-DECOR experiment and evidences for quark-gluon plasma in cosmic rays *Nucl. Instrum. Meth. A* **692** 228

[4] Zadeba E A et al. 2016 The registration system of the coordinate-tracking setup on the drift chambers *J. Phys. Conf. Ser.* **675** 032039.

[5] Bozhko N I et al. 1986 Drift chamber for the Serpukhov neutrino detector *Nucl. Instr. Meth. Phys. Res. A* **243** 388.

The optimization of nuclear power plants operation modes in emergency situations

A M Zagrebayev[1,2], A V Trifonenkov[1], R N Ramazanov[1]

[1]National Research Nuclear University MEPhI (Moscow Engineering Physics Institute), Department of Cybernetics, 115409, Russia, Moscow, Kashirskoe shosse, 31

[2]E-mail: AMZagrebayev@mephi.ru

Abstract. An emergency situations resulting in the necessity for temporary reactor trip may occur at the nuclear power plant while normal operating mode. The paper deals with some of the operation c aspects of nuclear power plant operation in emergency situations and during threatened period. The xenon poisoning causes limitations on the variety of statements of the problem of calculating characteristics of a set of optimal reactor power off controls. The article show a possibility and feasibility of new sets of optimization tasks for the operation of nuclear power plants under conditions of xenon poisoning in emergency circumstances.

1. Introduction

The present advancement of nuclear power allows building nuclear power plants in areas of unstable climatic conditions or with an increased risk of seismic activity, in remotely located areas where the nuclear power plant is the only source of energy, e.g. in the Arctic and Antarctic, as well as regions of political instability and terroristic threats [1]. Such power plants may undergo emergency situations. An example of an emergency is the earthquake and tsunami in Japan on March 11, 2011, which caused the accident at the Fukushima nuclear power plant. Thus, power plants can be exposed to emergencies, which cause abrupt nuclear reactor trips and their consequent idleness.

Almost all nuclear power reactors are currently thermal neutron reactors, for example, VVER, PWR, BWR type reactors. A significant physical feature of these reactors is the presence of xenon poisoning. Shutdowns at certain states of a reactor may cause long-term outages due to very high levels of xenon poisoning [2].

The process of xenon poisoning is described by the system of differential equations [3]

$$\begin{cases} \dfrac{\mathrm{d}x(t)}{\mathrm{d}t} = \lambda_{\mathrm{I}} i(t) - \lambda_{\mathrm{X}} x(t) - \sigma_{\mathrm{X}} \varphi x(t) \\ \dfrac{\mathrm{d}i(t)}{\mathrm{d}t} = \sigma_{\mathrm{X}} \varphi - \lambda_{\mathrm{I}} i(t) \end{cases} \tag{1}$$

The reactor control is impossible in case of exceeding the marginal xenon-135 concentration $x(t)$ level inside the reactor core [4, 5].

After reactor shutdown at τ, the minimal reactivity margin, needed for xenon poisoning compensation, changes. After the shutdown, the concentration of xenon first increases due to the decay of the iodine nuclei, and then decreases during the decay of xenon nuclei. The reactivity reserve $\Delta\rho \sim x(t)$ guarantees the criticality of the reactor until t_1 and after t_2 instances of reactor shutdown transient. At $t_1 < t < t_2$, the reactor power cannot be increased. In order for the reactor to be able to be put into operation at any time, the reactivity margin must compensate for the absorption of the maximum concentration by the xenon. This value depends on the neutron flux density change mode before the reactor shutdown. Reserving an additional amount of reactivity leads to a loss in the reactor's power production, since it means that the excessive fuel is not burned up.

The time when emergencies occur cannot be predicted precisely. A significant difference between natural emergencies and accidents caused by the human factor and the mechanism failure is, as a rule, a warning from the special services about the possibility of natural emergencies. There is usually a threatened period, which is declared beforehand, given by its duration and probability distribution for the event time. Different problems may be stated [3] on optimizing xenon poisoning of a nuclear reactor. Some of these problems are discussed below.

2. The task of optimizing the reactivity reserve in case of accidental shutdown of the power unit

Problem statement. Find reactivity margin value to reserve so that the average damage due to random reactor shutdown would be minimal.

The increase of the reactor power level is impossible if reactivity margin value is less than $\Delta\rho$, even if negative effects of the shutdown are eliminated after t_1. So the reactor power may be increased not before t_2.

In this case the reactor loses power production calculated as

$$\Delta Q_2 = W_{\max} \cdot (t_2 - t_1), \tag{2}$$

where $W_{\max}$ - design capacity of a nuclear reactor, MW.

Let the average value of power production loss of the reactor, which is shut down at random instance, be:

$$\Delta\overline{Q} = p_0 \cdot p(t_1) \cdot W_{\max} \cdot (t_2 - t_1), \tag{3}$$

where p_0 - the probability of unplanned reactor shutdown;

$p(t_1)$ - the probability of negative effects elimination after t_1.

Consider the case of the uniform probability distribution of the consequences elimination instance over $(0,t_2)$, that is $f(t) = \frac{1}{t_2}$. Then the probability of consequences elimination after t_1 is:

$$p(t_1) = \int_{t_1}^{t_2} f(t)dt = 1 - \frac{t_1}{t_2}.$$

In this case the average power production loss is calculated as

$$\Delta\overline{Q} = p_0 \cdot W_{\max} \cdot \frac{(t_2 - t_1)^2}{t_2} \tag{4}$$

In (3) t_1 and t_2 instances are functions of $\Delta\rho$. This dependence can be derived from the solution of equations, describing poisoning transient after the reactor shutdown:

$$\begin{aligned} \frac{dI}{dt} &= -\lambda_I \cdot I \\ \frac{dX}{dt} &= \lambda_I \cdot I - \lambda_x \cdot X \end{aligned} \tag{5}$$

given the initial conditions of steady state reactor poisoning:

$$I(0)=\frac{\gamma_I \Sigma_f \varphi_0}{\lambda_I}$$

$$X(0)=\frac{\gamma_I \Sigma_f \varphi_0}{\lambda_x+\sigma_x \varphi_0}$$

The solution of (5) given the initial conditions is:

$$X(t)=\frac{\gamma_I \Sigma_f \varphi_0}{\lambda_I}\{\frac{\lambda_I}{\lambda_x-\lambda_I}[\ell^{-\lambda_I\cdot t}-\ell^{-\lambda_x\cdot t}]+\frac{\lambda_I}{\lambda_x+\sigma_x\varphi_0}\ell^{-\lambda_x\cdot t}\} \tag{6}$$

The concentration of xenon is connected with the reactivity loss:

$$\rho_x=-\theta\frac{\sigma_x}{\Sigma_a}\cdot X(t) \tag{7}$$

where θ - thermal neutron utilization factor; Σ_a - macroscopic absorption cross-section, m^2.

The reactivity compensation for xenon poisoning is defined as

$$\Delta\rho=-\rho_x=\frac{\gamma_I \Sigma_f \varphi_0 \sigma_x}{\lambda_I \Sigma_a}\theta\{\frac{\lambda_I}{\lambda_x-\lambda_I}[\ell^{-\lambda_I\cdot t}-\ell^{-\lambda_x\cdot t}]+\frac{\lambda_I}{\lambda_x+\sigma_x\varphi_0}\ell^{-\lambda_x\cdot t}\} \tag{8}$$

The following estimations are valid for thermal reactors:

$\frac{\Sigma_f}{\Sigma_a}\approx 0{,}5$, $\sigma_x\cdot\varphi_0\approx 9\cdot 10^{-5}s^{-1}$, $\theta\approx 0{,}8$

The numerical estimation of reactivity margin for any outage duration is calculated as:

$$\Delta\rho=0{,}308\cdot(\ell^{-0{,}0756\cdot t}-\ell^{-0{,}1\cdot t})+0{,}019\cdot\ell^{-0{,}0756\cdot t} \tag{9}$$

It is clear, that $t_1(\Delta\rho)$ and $t_2(\Delta\rho)$ can be derived explicitly from (9). The average power production loss is the function of reactivity margin:

$$\Delta\overline{Q}=p_0\cdot W_{\max}\cdot\frac{(t_2(\Delta\rho)-t_1(\Delta\rho))^2}{t_2(\Delta\rho)} \tag{10}$$

The estimations show, that even with $p_0=1$ the power production loss given reactivity margin of 1% is never more then $\Delta\overline{Q}\approx 10^3\div 10^4 MW\cdot 24h$. At the same time, the power production loss for maintenance of such reactivity margin is estimated as $\approx 10^5 MW\cdot 24h$. Therefore, the reservation of additional reactivity margin is unnecessary. Another case is considered when unplanned shutdown causes serious economic circumstances. For example, if nuclear power plant is located in a remote area, its power production loss can't be compensated by other sources of electricity. Then, the cost of reserving additional amount of reactivity and the damage, caused by unplanned shutdown, are comparable. Thus, the following optimization problem is stated:

find the amount of reactivity to reserve, so that the average damage, caused by unplanned shutdown, would be minimal; the optimization criterion has the form:

$$\min_{\Delta\rho}\Delta Q=\frac{\Delta\rho}{a}+C\cdot p_0\cdot p(t_1)\cdot W_{\max}\cdot(t_2-t_1) \tag{11}$$

where C - the conversion factor of economic damage to the loss of power production. It can be defined as the division of ecomonic damage by the cost of electricity production.

Then the problem takes the following form:

$$\min_{\Delta\rho}\Delta Q=\frac{\Delta\rho}{a}+C\cdot p_0\cdot(1-\frac{t_1}{T})\cdot W_{\max}\cdot(t_2-t_1) \tag{12}$$

In fig. 1 the numerical estimations of power production loss, caused by the reservation of additional amount of reactivity, for various values of unplanned shutdown probability are plot. The calculations were performed given $a = 4.8 \cdot 10^{-6} \ \frac{1}{MW \cdot 24h}$, $W_{max} = 3200 \ MW$.

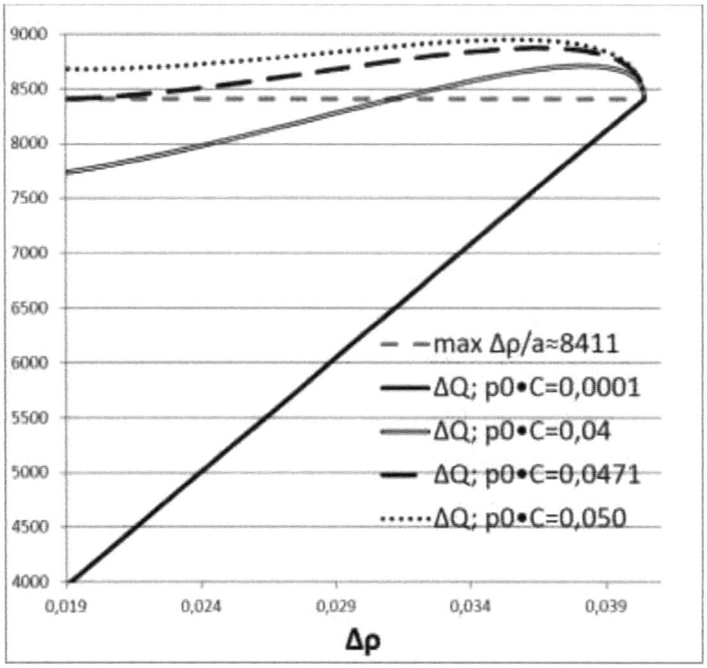

Figure 1. The dependence of estimated power production loss on the reactivity margin value

The figure shows, that optimal case is relay mode in relation to $C \cdot p_0$ coefficient. Physically, it means that optimal value corresponds to either absence of reactivity margin (for $C \cdot p_0 \leq 0.0471$), or having additional reactivity reserve for the reactor fast shutdown (for $C \cdot p_0 > 0.0471$).

3. Optimization of the regime for changing the power of a nuclear reactor during a threatened period in emergency circumstances

The solution of the problem above shows, that it is unnecessary to reserve the maximum amount of reactivity in regular conditions, but there are special cases, when unplanned shutdown causes serious economic circumstances. Earthquake is an example of a situation, when other sources of power can't compensate the loss. In such case the optimal solution is to reserve the amount of reactivity, sufficient for the reactor shutdown. This measure is extreme and in some cases another way can be considered. Indeed, when a short period of outage is assumed, which starts at random instance, described by some probability distribution, the problem of optimal power change mode can be stated.

Problem statement. Find the power change mode during emergency period to minimize the average amount of reactivity reserved for the compensation of maximum xenon concentration.

The optimization criterion has the form:

$$\min \int_0^T \Delta\rho_{max}(U(t)) f(\tau) d\tau \tag{13}$$

where $U(t)$ is the control during the emergency period.

$$U(t) = \begin{cases} 0 \leq \varphi(t) \leq \varphi_{max}, \ for\, 0 \leq t \prec \tau \\ 0, \quad for\, t \geq \tau \end{cases}$$

τ is random instance of emergency shutdown due to force-majeure circumstances, $0 \le \tau \le T$;
$f(\tau)$ is density of probability distribution for the instance of an emergency shutdown;
$\beta = \int_0^T f(\tau)d\tau$ is the probability of the reactor shutdown due to force-majeure circumstances.

Let $\varphi(t)$ be the solution of (1). Classic optimization theory, applied to the problem of control in case of xenon poisoning (optimal performance problem, minimal or maximal power production problem etc.), gives the optimal solution as the composition of following modes: $U = 0$, $U = \varphi_{\max}$, $U_{\text{classic}} = \varphi(t)$, $U_{\text{bound}} = \eta(t)$. The last two of them can be accurately approximated by step changes of power.

In this research the solution of the optimization problem (13) was carried out for $f(\tau)$ describing the uniform distribution and for step power change controls (see fig. 2): during the period Δt given, the power is kept at level $\alpha\varphi_{\max}$, then increased to maximum level $\varphi_{\max}$ during Δt and so on until shutdown is performed.

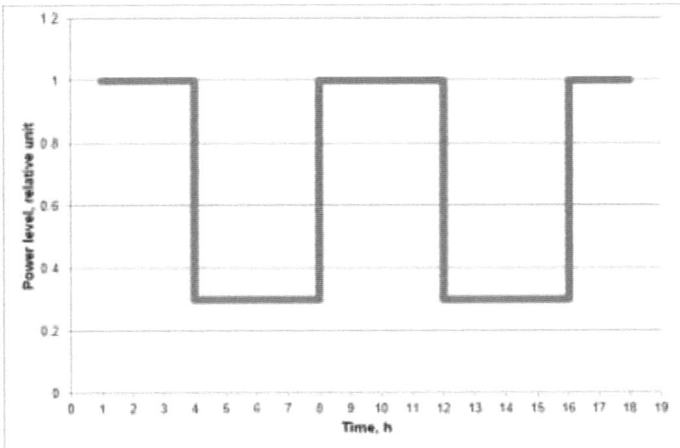

Figure 2. The power change mode

The table shows the values of the average reactivity reserve for various changes in capacity during the threatened period (T = 24 hours).

Table 1. The value of the average reserve of reactivity at various power reduction level (α).

α	Δt, hours						
	1	2	3	4	5	10	24
0	2.04	2.03	2.11	2.24	2.39	2.65	2.65
0.3	2.20	2.20	2.19	2.18	2.19	2.17	1.95
0.5	2.33	2.32	2.31	2.30	2.30	2.28	2.05
0.8	2.52	2.52	2.51	2.51	2.51	2.49	2.40

For example, if $\alpha = 0{,}3$, a stepwise change in power was investigated with a power reduction of up to 30% of the nominal value for different durations of the steps. When Δt = 1 hour switching control from $0{,}3\varphi_{\max}$ to $\varphi_{\max}$ and vice versa was carried out every hour, and Δt = 24 hours with only a single reduction in power up to $0{,}3\varphi_{\max}$. As follows from the table, it is this latter regime that requires the smallest average of the reactivity reserve for the possibility of escape from the iodine pit at any time $\Delta\rho_{\max} = 1{,}95\%$. In the standard approach, when the reserve of reactivity is reserved for the complete shutdown of the reactor immediately when the threat period is declared ($\alpha = 0, \Delta t = 24$ hours) the reactivity margin $\Delta\rho_{max} = 2{,}65\%$. Thus, the optimization effect is:

$$S = \frac{\Delta\rho_{max}(\alpha = 0, \Delta t = 24) - \Delta\rho_{max}(\alpha = 0{,}3, \Delta t = 24)}{\Delta\rho_{max}(\alpha = 0, \Delta t = 24)} = \frac{2{,}65 - 1{,}95}{1{,}95} = 36\%$$

The above results show that the optimization of the power change modes during the threatened period is appropriate and the optimal is a simple reduction of power to a certain level, the value of which depends on the specific parameters of the nuclear reactor.

4. The search for the interval of possible optimization of the power change mode in emergency circumstances

In the previous problem, the question was raised about the power change mode so as to minimize the average reactivity reserve. However, in practice, the reactivity reserve in the reactor can already be set, based on technological considerations. Then the task is to optimize the regime of capacity change during the threatened period within the existing limitations (on the amount of operational reactivity reserve and power level) [6].

From physical considerations it follows that the problem of optimal control in such a formulation with a restriction on xenon concentration has a solution not for any given duration of the control interval and not for any restriction on the xenon concentration. If the reactivity reserve is positive but does not compensate for the maximum poisoning, then at the beginning of the threatened period there is an interval of control duration values for which the solution of this optimal control problem does not exist because of the absence of solutions that satisfy the constraints.

Problem statement. Is there such a time period in which any management will lead to violation of the restrictions? Let the reactor operate at nominal (maximum possible) power and the reactivity margin for compensation of xenon poisoning is not. Then any decrease in power leads to the entry of the reactor into the iodine pit, that is, any optimization due to power change is impossible.

As it was mentioned above, the reactor shutdown at some set of states may cause a long-term reactor outage. That is why the condition for the values of both iodine-135 and xenon-135 at the end of process is introduced in order to limit the concentration of xenon-135 after the reactor shutdown. Finally, after the reactor shutdown occurs, there is an additional time span, reserved for the elimination of the accident's consequences, when the reactor can't operate.

Thus, the optimization problem statement conforms to the problem of optimal control on a limited interval with the limitation of xenon-135 concentration before shutdown and both iodine-135 and xenon-135 at the time of shutdown.

The threatened event time is random; it defines the duration of the process and the control function, which corresponds to the duration value. In practice, for the problem stated the optimal process can't be determined unless the period of control is given. Nevertheless, if the optimal control search problem is resolved in advance for any process duration value, the functions derived and their characteristics may be useful for applied control modes development.

In the time axis, originating simultaneously with the start of the reactor control process, there exists the time interval, containing instants, when the reactor shutdown doesn't satisfy the conditions. Further this interval is denoted by Δ_τ. The origin of this interval coincides with the origin of the time axis, the ending is defined by the parameters of the system and the limitations stated. The value of Δ_τ is difficult to obtain analytically, therefore numerical estimates are considered in the article.

One-group point model of a nuclear reactor [7] was used in the research. For the case of the nuclear reactor after having been shut down, the system of differential equations takes the following form

$$\begin{cases} \dfrac{\mathrm{d}x(t)}{\mathrm{d}t} = \lambda_I i(t) - \lambda_X x(t) \\ \dfrac{\mathrm{d}i(t)}{\mathrm{d}t} = -\lambda_I i(t) \end{cases} \tag{14}$$

Initial conditions: $i(\tau) = i^0$, $x(\tau) = x^0$.

Two ways of estimating the length of the interval were considered. The first way corresponds with the lower estimation.

Let t_0 be the origin of the threatened period and t_1 be the instant of the reactor shutdown. In these terms, if $t > t_1$ then $\varphi(t) \equiv 0$ and the transients are described by (14).

Let $\mathcal{I}$ be the set of admissible control functions $\varphi(t)$, such that $0 \le \varphi(t) \le \varphi_{\max}$, where $t_0 \le t \le t_1$ is the time span of the control process. The set of states of the model (1) is represented by the quadrant $Q = \{(i,x) \mid i \ge 0, x \ge 0\}$ of the phase plane Oix.

Let A be the set of probable model states at time τ, given $i(t_0) = i_0, x(t_0) = x_0$ for any $\varphi(t) \in \mathcal{I}$. Let B be the set of acceptable states at $t = t_1$ that is the end of the control process. In these terms, if $A \cap B = \{\varnothing\}$ at some instant τ, then, if $t_1 = \tau$, the problem is inconsistent. This is to be proved by the following.

To investigate the properties of the set B, consider the function $F(i,x)$ over the quadrant Q; the function is defined by (14) solution properties as follows. Let $i(t), x(t)$ be the solution of (14) given initial conditions $i(t_1) = i_1, x(t_1) = x_1$, where $(i_1, x_1) \in Q$ and $t \ge t_1$. By construction, $F(i_1, x_1) = \max\limits_{t \in [t_1, +\infty)} x(t)$. Consequently, $B = \{(i,x) \in Q \mid F(i,x) \le x_{\text{lim}}\}$.

The equation describing the phase curve for the system (2) has the form:

$$x = \frac{\lambda_I i_m}{\lambda_I - \lambda_X}\left(\left(\frac{i}{i_m}\right)^{\frac{\lambda_X}{\lambda_I}} - \frac{i}{i_m}\right), \text{ with } 0 \le i \le i_m .$$

Let $f(i)$ be the right-hand member of this equation; note also that $f(i)$ is convex upwards at $0 < i < i_m$. Then the equiscalar lines of $F(i,x)$ have the form of $x = L(i)$, where

$$L(i) = \begin{cases} f(i), i^* \le i \le i_m \\ f(i^*), 0 \le i < i^* \end{cases} \text{ and } i^* = i_m \left(\frac{\lambda_X}{\lambda_I}\right)^{\frac{\lambda_I}{\lambda_I - \lambda_X}}$$ - the maximum point of $f(i)$ (see figure 3 below).

It is clear that the value of $F(i,x)$ over the equiscalar line equals the maximal value of $f(i^*)$.

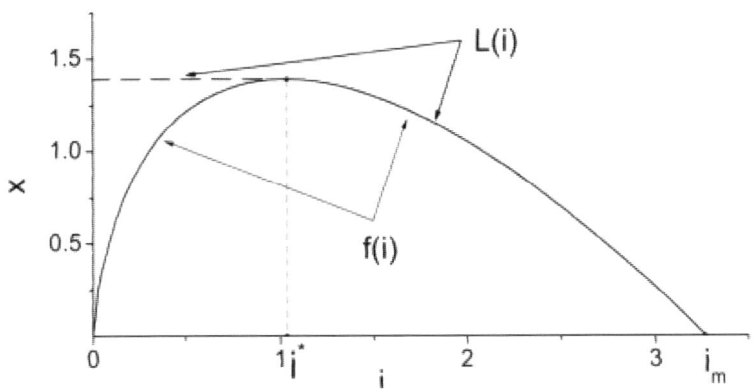

Figure 3. The form of the equiscalar line $x = L(i)$

Consequently, the function $L(i)$ is also convex upwards at $0 < i < i_m$; furthermore, the set B is also convex. The boundary of the set B at $i > 0$ and $x > 0$ is defined by the equiscalar line $x = L(i)$, corresponding to the value $F(i,x) = x_{\text{lim}}$.

The set B is closed, and, if $(i_0, x_0) \notin B$, then there exists a circular area $U_0 = \{(i,x) \mid \sqrt{|i-i_0|^2 + |x-x_0|^2} \le a\}$ $(a > 0)$, such that $U_0 \cap B = \{\varnothing\}$. So, the estimation [8]

$(i(t), x(t)) \in U_0$ where $t_0 \le t \le a/M$ for any $\varphi(t) \in \Phi$, given

$$M = \sup_D \sqrt{(\lambda_I i - (\lambda_X + \sigma_X \varphi)x)^2 + (\sigma_X \varphi - \lambda_I i)^2}$$

and $D = \{(i, x, \varphi) \in R^3 \mid (i,x) \in U_0,\ 0 \le \varphi \le 1\}$,

which follows from the boundedness of right-hand members of (1) over the set D, results in $\Delta_\tau > a/M > 0$, because for any $t_0 \le t \le a/M$ the set $A \subset U_0$ and, consequently, $A \cap B = \{\varnothing\}$.

More precise lower estimation of Δ_τ may be obtained by adjusting the shape of sets A and B. To simplify the computation procedure convex polygons are considered; they are denoted by A_c and B_c, and contain sets A and B respectively. Thus, if $A_c \cap B_c = \{\varnothing\}$, then sets A and B do not intersect either.

The conditions, considered in this article, implying that the initial state (i_0, x_0) coincides with the equilibrium point of the system (1) given $\varphi = \varphi_{\max}$ and $(i_0, x_0) \notin B$, then the construction of the set B_c may be based on the following reasoning. In the coordinate system $Oixh$ the tangential plane defined by the equation $h = F(i,x)$ at the point $(i_0, x_0, F(i_0, x_0))$ is considered. The Oix phase plane projection of the intersection line between the tangential plane and the plane $h = x_{\lim}$ may be used as a part of the boundary of the set B_c (line Γ in figure 4).

The sides of the polygon A_c, estimating the set A, are formed by K estimations for the solutions of (1) denoted by $i(t), x(t)$ given $i(t_0) = i_0, x(t_0) = x_0$ and any $\varphi(t) \in \Phi$

$s_k(t) \le \alpha_k\, i(t) + \beta_k\, x(t)$, $|\alpha_k| + |\beta_k| > 0$, for $t_0 \le t \le t_1$, $k = 1, \ldots, K$.

The estimations of such type are obtained by the integration of differential inequalities, which follow from (1) given the limitation $0 \le \varphi(t) \le \varphi_{\max}$, signs of summands in right-hand members of equations, as well as signs of coefficients α_k and β_k for the specific estimation line. For example,

1) $s_1(t) = i_0 \exp(-\lambda_I (t - t_0))$, $\alpha_1 = 1, \beta_1 = 0$;

2) $s_2(t) = (x_0 - c_0)\exp(-(\lambda_X + \Lambda)(t - t_0)) + c_0 \exp(-\lambda_I (t - t_0))$, where $c_0 = \dfrac{\lambda_I i_0}{\lambda_X + \Lambda - \lambda_I}$,

$\Lambda = \sigma_X \varphi_{\max}$, $\alpha_2 = 0, \beta_2 = 1$;

3) $s_3(t) = i_0 + x_0 - (\lambda_X + \Lambda)\big((x_0 - d_0 - \Lambda/\lambda_X)(1 - \exp(-\lambda_X (t - t_0)))/\lambda_X +$

$+ \Lambda (t - t_0)/\lambda_X + d_0 (1 - \exp(-\lambda_I (t - t_0)))/\lambda_I\big)$, where $d_0 = \dfrac{\lambda_I i_0 - \Lambda}{\lambda_X - \lambda_I}$, $\alpha_3 = 1, \beta_3 = 1$;

etc. The choice of the count K of estimations and their coefficients α_k and β_k may be varied to adjust the set A_c (see figure 4 below).

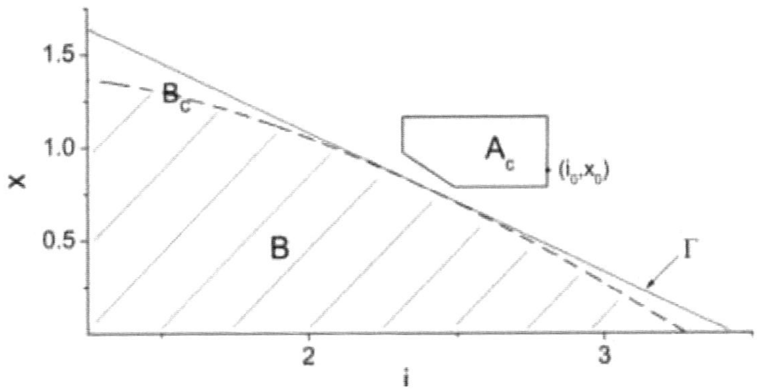

Figure 4. Approximations of sets A and B; the intersection line Γ

Then the relative position of A_c and B_c is analyzed for various values of τ and the point of time is estimated when the condition $A_c \cap B_c = \{\varnothing\}$ is violated. The maximal duration of the time interval $t_0 \le \tau < t_0 + \Delta_\tau$, where each point τ satisfies $A_c \cap B_c = \{\varnothing\}$, defines the lower estimation for Δ_τ.

The upper estimation of the length of the interval was based on the solution of the problem of optimal speed reactor power decrease with limited xenon concentration. The statement of this problem and analytical principles of the solution construction are stated in [9]. The duration of the numerically calculated process is the upper estimation for the interval. The duration of the process as the function of the xenon limitation value was plot (see figure 6 below).

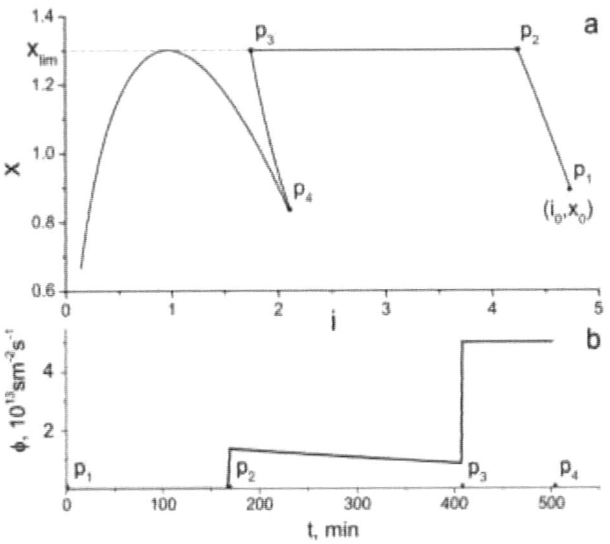

Figure 5. (a) The example of a transient, representing the optimal speed reactor power decrease, over the phase plane, (b) optimal control

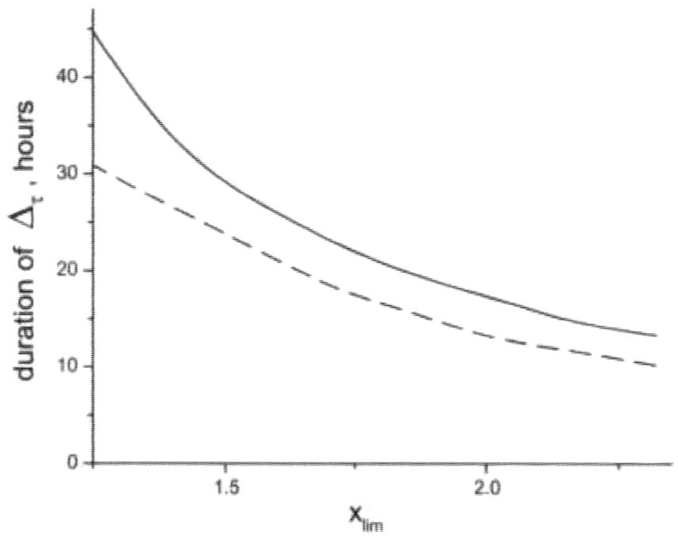

Figure 6. The estimations of the duration of segment Δ_τ as the functions of the limitation value for the concentration of xenon: solid and dashed lines for upper and lower estimations respectively

As results of the research carried two ways of estimation of the time interval observed were developed and applied and the dependencies of the estimations as functions of xenon limitation value were plot. The results obtained may be used in further research of the improvements to the problem statement considered in this article.

5. The search of the best time shift for the quickest reactor recovery to the nominal power after the reactor trip

In the previously considered problems, it was about optimizing the operating mode under emergency circumstances of one power unit. However, as a rule, a nuclear power plant consists of several power units whose maneuvering capabilities may differ. For example, power units have different reactivity reserves, since they were overloaded at different times.

It proposes the formulation and solution of the optimization task [10] of choosing the best time shift between the refuelling of nuclear reactors, which allows the maximum rapidity to reach the rated power after the power plant is shut down.

For simplicity, the NPP consists of two VVER-type reactors; the reactivity margin of each of them for compensation of xenon poisoning depends on the time elapsed since the beginning of the campaign. T_0 is the operating time of the reactor between the overloads, τ is the time shift between the refuelling of the two power units, Δt is the time required for refuelling. The reactivity of the reactors of the two power units is shown in figure 7 below.

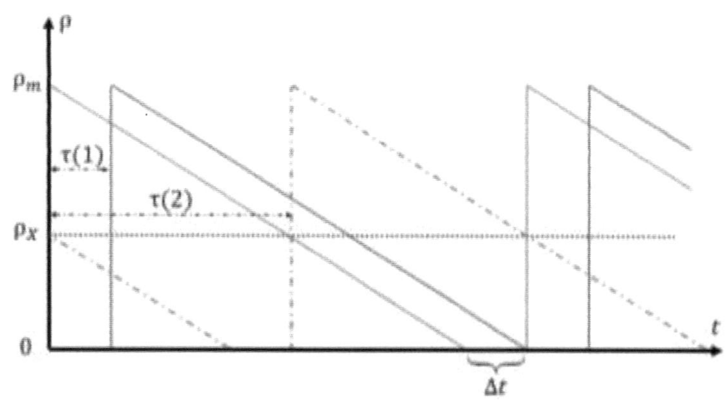

Figure 7. Dependence of reactivity on time within the campaign

With brief shifts τ(1) between refuelling after unscheduled trips of both reactors, a period exists when neither can restore nominal power, since both stay in iodine pit. For the rest of time both reactors can restore nominal power in no time. On the contrary, with large time shifts τ(2) one of the reactors at a certain point would possess reactivity margin to compensate the iodine pit, when the other still stays in the iodine pit.

Thus, the shift of τ significantly affects the maneuverability of nuclear power plants. Let us find the time shift after the refuelling should be in the reactors, in order to increase the power to the nominal level in the minimum time after the end of the threatened period.

If the reactor stays at such a campaign stage, that it is impossible or undesirable to handle the control rods, to get out of the iodine pit one should alter boric acid concentration in the active zone (letdown). Let C_0 is the concentration of boric acid at the beginning of the transient process, G - letdown rate, V - the volume of the primary circuit. Then the concentration of boric acid in the case of make-up by distillate will be calculated by the formula: $C(t) = C_0 \exp(-Gt/V)$. Reactivity margin due to boric acid presence in the coolant is proportional to its concentration. If it is required to release the reactivity margin $\Delta\rho_x$, the concentration of boric acid should be reduced in accordance with the following equation:

$$\rho(T) - \rho(t) = \rho(T)\left[1 - \exp(-Gt/V)\right] = \Delta\rho_x$$

Let the boric acid concentration time dependence within single campaign be approximated with enough accuracy by linear function. Thus, if the reactor was stopped at the time of the T campaign, then it takes time for it to exit the iodine pit:

$$t = -\frac{V}{G}\ln\left(1 - \frac{\Delta\rho_x}{\rho_0 - \alpha t}\right)$$

The further the reactor is from the beginning of the campaign, the more time it will take to purge it to provide a reserve of reactivity sufficient to compensate for xenon poisoning. If the reactor trip occurred in the interval (0; τ), then the reactivity reserve is greater for the first reactor. Therefore, it must be the first to reach power. If the stop has occurred in the interval (τ; T_0), then the second reactor.

Since the moment of trip is random, we can speak about average letdown time for starting one reactor at the power level depending on the time shift. Suppose that the moments of occurrence of emergency circumstances within the campaign are equally probable. The average time to power output will be expressed by the ratio:

$$\bar{t} = -\frac{V}{GT_0}\left[\int_0^{\tau}\ln\left(1 - \frac{\Delta\rho_x}{\rho_0 - \alpha T}\right)dT + \int_{\tau}^{T_0}\ln\left(1 - \frac{\Delta\rho_x}{\rho_0 - \alpha(T - \tau)}\right)dT\right]$$

If we set the optimization task of finding the time shift τ so that in a minimum time after the cancellation of the threatened period, at least one power unit could reach its rated power. Then taking the derivative of average time with respect to the parameter τ, we obtain for the case of two identical reactors τ = 50% T_0 (see figure 8 below). If we set the task of finding the time shift τ to minimize the time to power output after the end of the threatened period of the entire NPP, then in this situation the minimum will be reached at 0 and 100%, that is, in the absence of a shift (see figure 9 below).

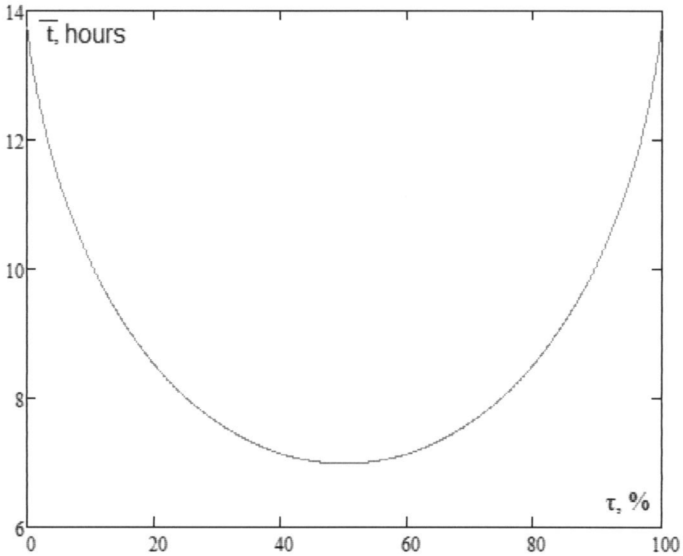

Figure 8. The average time taken to output one unit (in hours), depending on the time shift (in percent of the campaign)

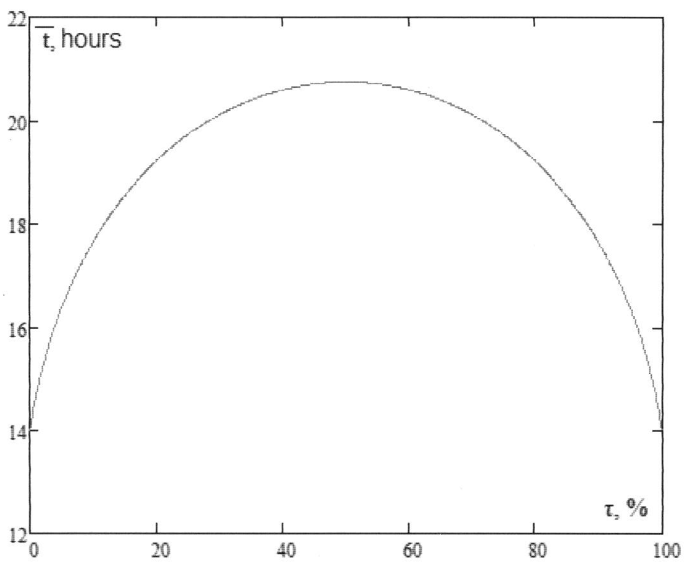

Figure 9. The average time taken to output all units (in hours), depending on the time shift (in percent of the campaign)

Thus, it has been shown that the presence of a time shift between reactor overloads at nuclear power plants allows one unit to be produced as quickly as possible in the case of force majeure, but not all reactors can be simultaneously outputted to power. In the general, for NPP consisting of N reactors the optimum shift is 1/N of the reactor campaign.

6. Conclusions

The problems discussed above show that there is a possibility and feasibility of new sets of optimization tasks for the operation of nuclear power plants under conditions of xenon poisoning in emergency circumstances.

References

[1] Zagrebayev A M, Ovsyannikova N V and Sadchikov S M 2012 Optimization of a mode of

change of a nuclear reactor capacity during the threatened period under force-majeur circumstances *Modern problems of science and education* **4** 1–7

[2] Zagrebayev A M, Ovsyannikova N V and Sadchikov S M 2012 The average loss of energy production at a random stop of a reactor with a limited operational reserve of reactivity *Natural and technical sciences* **58** 418–22

[3] Rudik A P 1979 *Optimization of reactor physical characteristics* (Moscow: Atomizdat)

[4] Dollezhal' N A and Emel'yanov I Ya 1980 *Channel Nuclear Power Reactor* (Moscow: Atomizdat)

[5] Ovchinnikov F Ya, Golybev L I, Dobrynin V D, Klochkov D I, Semenov V V and Tsybenko V I 1979 *Operating conditions of water-moderated-water-cooled power reactors* (Moscow: Atomizdat)

[6] Trifonenkov A V and Trifonenkov V P 2017 On a distinctive feature of problems of calculating time-average characteristics of nuclear reactor optimal control sets *IOP Conf. Series: Journal of Physics: Conf. Series* **788** 1–6

[7] Zweifel P 1997 *Reactor Physics* (Moscow: Atomizdat)

[8] Pontryagin L S 1974 *Ordinary differential equations* (Moscow: Nauka)

[9] Rudik A P 1971 *Nuclear reactors and Pontryagin's maximum principle* (Moscow: Atomizdat)

[10] Zagrebayev A M and Ramazanov R N 2016 Searching for the best time shift for the fastest reactor recovery to nominal power after reactor trip *Trudy MAI* **91** 1–9

IOP Conf. Series: Journal of Physics: Conf. Series **945** (2017) 012029 doi:10.1088/1742-6596/945/1/012029

The barometric effect in the intensity of near-horizontal cosmic ray muons according to the data of the coordinate-tracking detector DECOR

D V Zaitseva and R P Kokoulin

National Research Nuclear University MEPhI (Moscow Engineering Physics Institute) Kashirskoe Shosse 31, 115409 Moscow, Russia

E-mail: zaytseva2@mail.ru

Abstract. Experimental data on cosmic ray muons obtained with the coordinate-tracking detector DECOR in 2012 – 2017 are analyzed. The rate of the events exhibits variations significantly exceeding the statistical errors. It is shown that these variations are related to changes in the meteorological conditions (atmospheric pressure and air temperature). Barometric coefficient for near-horizontal muons has been estimated.

1. Introduction

In the frequency of the events formed as a result of interactions of cosmic rays in the atmosphere and recorded at the surface of the Earth, variations are observed that are related to the state of the atmosphere (atmospheric pressure, altitude distribution of air temperature and, respectively, air density, moisture content, etc.) [1].

An increase in the atmospheric pressure leads to an increase in the mass of the matter above the observation point and to a decrease in the intensity of muons due to the absorption of the less energetic particles in the atmosphere. This phenomenon is called the barometric effect and is approximately described by the introduction of the so-called barometric coefficient. For muons, the barometric effect decreases rapidly with increasing energy. In present work, the barometric effect for near-horizontal muons is studied. In contrast to the near-vertical muons, the particles in the investigated range of angles pass through the atmosphere a hugely larger path (geometrically ~ 400 km, and by the mass of matter ~ 20 kg/cm^2) [2]. Earlier, variations in the intensity of near-horizontal muons were not studied.

2. Experimental data

Experimental data on cosmic ray muons obtained with the coordinate-tracking detector DECOR [3] in 2012 – 2017 are analyzed. DECOR is designed for studying multi-particle events and single muons at large zenith angles. The detector is deployed around the Cherenkov water calorimeter NEVOD. DECOR consists of 8 supermodules (SMs) with total area of ~ 70 m^2. Every SM includes 8 vertical streamer tube chamber planes. The events reconstructed from the data of SMs located in the opposite short galleries are analyzed. If the track segments registered in two SMs agree within a cone of 5°, it is supposed that both SMs were crossed by the same particle (a muon with energy of at least 7 GeV). Such events correspond to zenith angles from 85° to the horizon. The average zenith angle is 87.5°; the calculated average muon energy is about 100 GeV.

After preliminary analysis, 916 data sets (runs) with duration from 10 to 40 hours of «live time» were selected. The total live time amounted to 28508 hours. About 3% of available data were excluded from the consideration because of some fault conditions disclosed in them or runs were lasting less than 10 hours. The rate of the events exhibits variations significantly exceeding the statistical errors. It is naturally to assume that these variations are related to changes in the meteorological conditions (atmospheric pressure and air temperature).

The frequency of events selected during one of the periods of the measurements is represented by the points in the figure 1 (each point corresponds to a separate set); the line represents the atmospheric pressure values averaged over the set time. A negative correlation of the frequency of events with pressure is clearly noticeable: the extremely high values of atmospheric pressure correspond to the minimum values of the event frequency; on the other hand, an increased rate of events is observed, as a rule, during periods of a sharp drop in the atmospheric pressure.

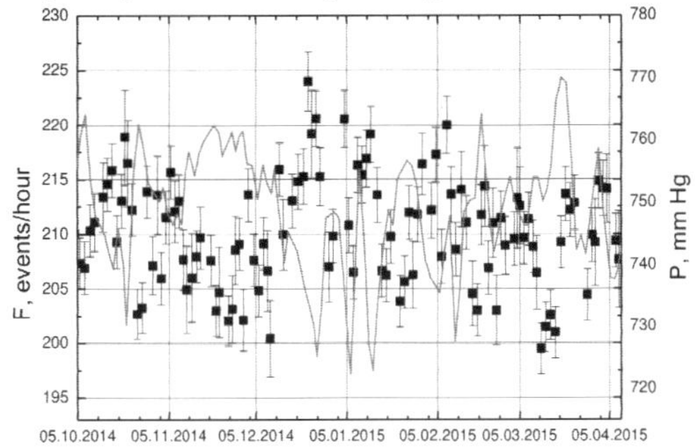

Figure 1. Changes in the frequency of events and pressure over the period October 2014 - April 2015.

Values of atmospheric pressure were obtained with the help of a sensor, located near the installation (in the NEVOD building). The sensor measures the pressure every 22 seconds; from these data, the average pressure per set was calculated. In the sensor data, time intervals corresponding to each data set were found, and pressure averaging over these intervals was performed. Thus, for each data set, the mean atmospheric pressure for the NEVOD sensor was determined. The average pressure for all sets of three series of observations was also calculated (see table 1).

3. Barometric coefficient

A barometric coefficient is used to quantify the barometric effect. Empirically, it can be determined by a linear correlation between the data on the intensity of events with near-horizontal muons and atmospheric pressure data. The correlation between the frequency of selected events and atmospheric pressure for all series of observations is shown in the figure 2.

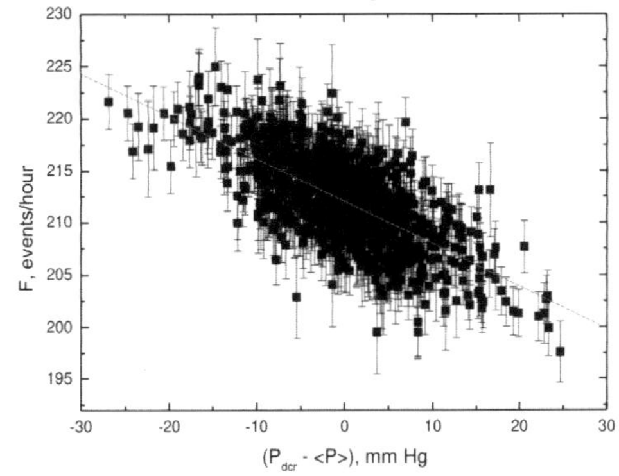

Figure 2. Correlation of event frequency and pressure for three series of observations. The line is the linear approximation (1).

Linear approximation of the obtained dependence by the method of least squares (the line in the figure) allows to obtain an estimate of the barometric coefficient for a given class of events. The line in the figure is the result of applying linear regression taking into account the statistical weights of the individual sets. The slope of the regression line for this relationship corresponds to the barometric coefficient.

The frequency of the selected events is approximated by a function in the form:

$$F_i = F_0(1+\beta_p \Delta P) \ ; \ \Delta P = P - < P > , \qquad (1)$$

where P is the current pressure value for a given set, $<P>$ is the pressure averaged over the whole series of measurements, and β_P is the barometric coefficient.

The mean values of the atmospheric pressure over the experimental period and the barometric coefficient estimates for individual series, as well as for the entire observation period, are presented in the table 1.

Table 1. Summary results of estimating the barometric coefficient for the intensity of the near-horizontal muons of cosmic rays from the data of the present experiment.

Series	I	II	III	I - III
$<P>$, mm Hg	746.3	747.0	746.1	746.6
β_P, %/mm Hg	-(0.169±0.013)	-(0.201±0.008)	-(0.178±0.012)	-(0.192±0.005)

As can be seen from the table, the barometric coefficient estimates for different series are in satisfactory agreement with each other. It should be noted that the obtained estimates include not only the actual changes in the intensity of the near-horizontal muons, but also the apparatus component related with possible changes in the recording and event selection efficiency with changes in the atmospheric pressure. According to our estimates, the apparatus contribution to the barometric coefficient is about - (0.024±0.002) % / mm Hg. Thus, the final value of the barometric coefficient for near-horizontal cosmic ray muons with energy more than 7 GeV can be evaluated as - (0.168±0.005stat±0.002syst) % /mm Hg.

Acknowledgments
The work was performed at the experimental complex NEVOD with the support of RF Ministry of Education and Science (contract RMEFI59114X0002 and MEPhI Academic Excellence Project).

References

[1] Dorman L I 1972 Meteorological effects in cosmic rays (Moscow: Nauka) p 211
[2] Amineva T P et al. 1975 Investigation of ultra-high-energy muons, (Moscow: Nauka) [in Russian] p 216
[3] Barbashina N S, Ezubchenko A A, Kokoulin R P et al. 2000 A coordinate detector for studying horizontal fluxes of cosmic ray (Instruments and Experimental Techniques vol 43), pp 743-746

IOP Conf. Series: Journal of Physics: Conf. Series **945** (2017) 012030 doi:10.1088/1742-6596/945/1/012030

Hardware-software complex of in-depth studies of hexapod walking robot

V Kh Ankudinov, A V Maksimov, S O Starkov

Kaluga branch of Bauman Moscow State Technical University, Russia, Kaluga, Obninsk State Institute of Nuclear Engeneering – a branch of National Research Nuclear University MEPhI, Russia, Obninsk

vladislav.ankudinov@gmail.com, av_maximov@bk.ru

Abstract. A hardware-software complex of in-depth studies of hexapod walking robot including the hexapod mockup and its simulator is described. Possible applications of the complex for the study of statically stable hexapod gaits and obstacle clearance algorithms are considered. An important feature of the simulator is the ability to work in real time that allows you to use it in the control loop.

1. Introduction

To work in areas that are inaccessible or dangerous for humans, it is necessary to develop new types of robots that can autonomously or by the operator's command avoid obstacles, overcome blockages, climb stairs, and so on. One of the possible ways to solve this problem is the use of a six-legged walking robot – hexapod, Figure 1.

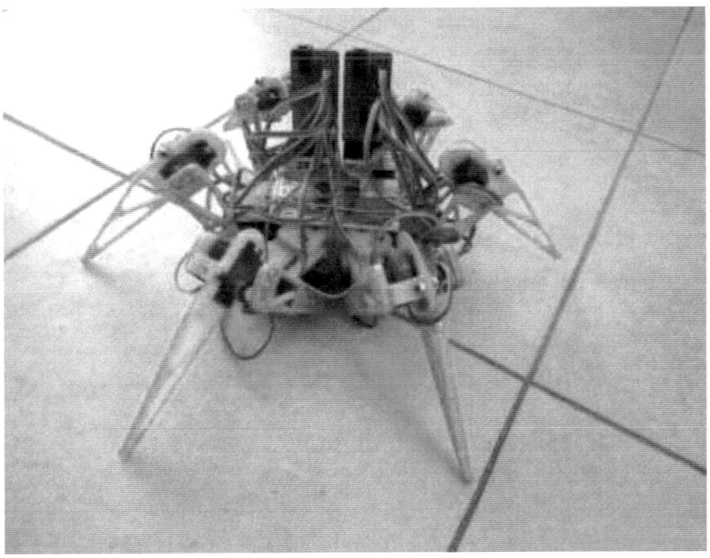

Figure 1. Hexapod.

Figure 1 shows the "Snowflake" hexapod made for research purposes. The robot parts are designed in the SolidWorks system and printed on a 3D printer made of ABS plastic. As it is known, for the static stability of the robot it is enough constant reliance on three legs. Therefore, for six-legged machines it is possible to synthesize a statically stable gait in any of its phases and the ability to maintain this stability even with a partial failure of one or even two legs. An important feature of hexapods is the flexibility in the formation of the initial (starting) orientation and the possibility of maintaining this orientation in the process of movement, even when obstacles are overcome, regardless of the direction of motion. This can be useful when monitoring (video inspection) of the environment or when transporting goods sensitive to impacts or turns in space.

An analysis of available publications, for example [1-3], shows that currently there are no techniques for designing the considered class of mobile robots that are acceptable for practical application, there are no techniques for designing control systems for their movement. In this connection, there is a need to develop means for studying hexapods with the subsequent development of the main provisions of the engineering method of controlling their movement.

2. Hardware-software complex

The hardware and software complex includes a hexapod mockup and its simulator. The structural diagram of the complex is shown in Figure 2.

The simulator is intended for verification of kinematic models and development of control algorithms. Gives a good repeatability of experiments independent of the servo drives, the condition of the communication channel, temperature and other external factors. In the simulator it is easier to debug the control algorithms, because we have full information about the simulated objects and we can control the simulation progress, incl. stop it. Experiments, successfully conducted in the simulator are then repeated using the physical mockup of the hexapod.

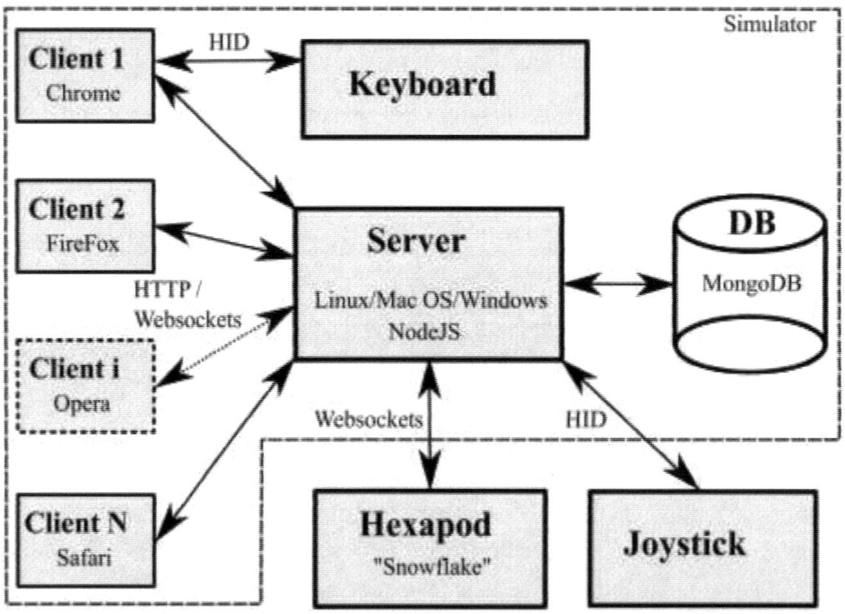

Figure 2. Hardware-software complex.

As you can see in Figure 2, the simulator is based on web technologies. The client does not need to install any software, it is enough to have a modern web browser with support for WebGL, WebSockets and Web Workers. The server is written in Node.js, MongoDB is used as the database. Deployment of the server is possible on most popular operating systems today, including Windows, Linux, Mac OS.

Experiments can be easily parallelized. Calculations are made on the client side, i.e. The more customers - the more processing power.

As can be seen from Figure 1, the hexapod "Snowflake" has a radial symmetry and is made according to an insectomorphic scheme. Each of his six legs has three degrees of freedom - one in the horizontal plane and two in the vertical plane.

Currently, the hexapod is controlled through a simulator. The researcher is given two types of interfaces: high-level and low-level. The low-level interface gives access to the rotation angle of each of the 18 servos and can be used to synthesize various gaits by the researcher. A high-level interface is provided to the operator for solving applied problems and investigating intelligent control algorithms: path selection, obstacle clearance, etc.

In the hexapod there are 18 servo drives, the control of the operator behind each of them in real time seems inconvenient and impossible. The joystick is a familiar and convenient controller, therefore, based on the high-level API, methods for managing the hexapod using a joystick and a conventional keyboard have been developed and tested, figure 3. In addition, the user is provided with a low-level API for developing his own methods and techniques of control.

Figure 3. Joystick control.

On the basis of the software and hardware complex, a number of experiments were carried out. In particular, the gait of the physical sample of the hexapod on a smooth surface is adjusted. And in the simulator, experiments were conducted to overcome various obstacles.

At this stage hexapod does not have a system of technical vision because of its great technical complexity. The lack of a visual system is easily replenished by a simulator, with which you can simulate the state of the surrounding space, the surface of the robot's movement. So when simulating the overcoming of an obstacle, for example, moving a robot along the steps, the operator is able to fully control the position of its hull, which can be useful when working out algorithms for transporting goods that are sensitive to position in space or impacts. And this kind of movement is an important feature of walking robots, which is inaccessible to wheeled and caterpillar vehicles.

3. Main results

In the course of research using the software and hardware complex, the following results were obtained:

- A kinematic model of hexapod based on homogeneous matrices was created and verified [4]. The model is convenient in engineering practice.
- A kinematic model of the hexapod in dual quaternions is created. This model has more functionality and computational efficiency.
- Various gaits were synthesized (including one-legged and three-legged). Three-legged gait is the fastest of statically stable gaits. One-leg gait is the most stable. In a one-legged gait, the hexapod has the largest carrying capacity.
- Developed and debugged methods for controlling the motion of the hexapod with a joystick and keyboard, which significantly reduces the complexity of training attendants.

Currently, work is underway to create an algorithm for rearranging legs while moving a hexapod through movable obstacles of arbitrary shape.

4. Prospects of the development of the complex

- Due to easy parallelization of experiments, good repeatability and centralized storage of data, the simulator can be adapted for carrying out computationally capacious experiments, such as searching for optimal gaits using genetic algorithms or neural networks.
- When deploying a simulator in a telecommunications network, it becomes possible to attract a large number of operators simultaneously for large-scale research and experiments, to increase the speed of rescue operations, and to conduct various contests and competitions.

Acknowledgements

This work was supported by Competitiveness Program of National Research Nuclear University MEPhI.

References

[1] Garcia-Lopez M C, Gorrostieta-Hurtado E, Vargas-Soto E, Ramos-Arreguin J M, Sotomayor-Olmedo A and Moya Morales J C 2012 Kinematic analysis for trajectory generation in one leg of a hexapod robot *The 2012 Iberoamerican Conf. on Electronics Engineering and Computer Science (Procedia Technology* vol 3*)* ed C A Gutierrez et al pp 342-350

[2] Mahdi M 2013 Analitical workspace, kinematics and foot force based stability of hexapod walking robots *Degree of Doctor of Philosophy in Mechanical Engineering* (Worchester: Worchester Polytechnic Institute)

[3] Hoerger M 2014 Real Time Stabilisation for hexapod robots using task-space constraints *Online publications of the University of Stuttgart* (Stottgart: Systems Department of Machine Learning and Robotics University of Stuttgart)

[4] Ankudinov V Kh, Maksimov A V 2016 Kinematics model of hexapod. Part 1. Matrix models *Electromagnetic Waves and Electronic Systems* **21** 56-65

IOP Conf. Series: Journal of Physics: Conf. Series **945** (2017) 012031 doi:10.1088/1742-6596/945/1/012031

Approximation of the criticality margin of WWR-c reactor using artificial neuron networks

I Belyavtsev[1], **D Legchikov**[1], **S Starkov**[1], **V Kolesov**[1] **and E Nikulin**[1]

[1] Obninsk State Institute of Nuclear Engeneering — a branch of National Research Nuclear University MEPhI, Studgorodok, 1, Obninsk, Kaluga Region, Russia, 249040

E-mail: djbelyak@gmail.com

Abstract. Two artificial neural networks approximating the criticality margin of the WWR-c reactor (based on model and calculated data) were created and trained. The resulting neural networks realize the correct approximation, have high accuracy, and also high speed of operation. Using the obtained artificial neural networks can be applied to accelerate the preliminary calculations of the state of the reactor.

1. Formulation of the problem

There is an experimental nuclear facility WWR-c in the Karpov Institute of Physical Chemistry in the city of Obninsk. The nuclear facility was ccommissioning into operation in 1964. WWR-c is a heterogeneous water-water research reactor, specialized for conducting a wide range of experiments in the field of radiation chemistry, structural and material research, activation analysis, neutron doping of semiconductors, etc. [1]

Since 1980, the productions of radionuclides for medical purpose and radiopharmaceuticals based on them have been developing on the reactor. In connection with the success of the development of the production and the convenient geographical location in 1986, it was decided to improve the reactor. [1]

To improve reactor parameters and the efficiency of radionuclide production (^{99}Mo, ^{131}I etc.)the precision neutron-physical model for active zones, reflectors and control systems was created in 2011. A geometry of all fuel elements (fuel, shell, water gap, temperature of each component), a change of isotopic composition as a function of burn-up, control systems, reflectors, experimental channels and structures were taken to account during the simulation. The precise model were verified for calculation of the criticality margin of WWR-c reactor. [2]

The precision model is based on the Monte Carlo method, which makes it possible to achieve great accuracy in the modeling of extremely complex physical processes in the active zone of the reactor. However, for carrying out a correct calculation, a large number of computational experiments are required, and, consequently, a large amount of computer time. Calculation of one state of the active zone takes about 8 hours.

Obviously, with short reactor campaigns (100 hours), the research personnel of the reactor have little time for calculating the modernized state of the reactor. Thus, our task is to create an approximation of the precision model for preliminary calculations of the criticality margin of reactor. Approximation should satisfy the following requirements: high speed of operation and sufficient accuracy.

Let consider parameters of the precise model which are changing from campaign to campaign. This is the percentage of fuel burn-out in each fuel assembly and the position of the control systems. Other model parameters from campaign to campaign remain unchanged. In this case, the task is reduced to approximating the criticality margin of the reactor, depending on the fuel burn-out and the position of the control system.

2. Experiment #1: Model Data Approximation

To construct the approximation, a generalized approximation theorem was applied. According to this theorem it is possible to obtain an arbitrarily exact approximation of any continuous function of several variables, using the operations of addition and multiplication by a number, a superposition of functions, linear functions, and also one arbitrary continuous nonlinear function of one variable. [3] Since these operations are fully implemented by an artificial neural network with one nonlinear formal neuron, an artificial neural network can be used to construct the desired approximation.

To build the approximation based on an artificial neural network, we used the TensorFlow library. The TensorFlow library is the most popular library for machine learning. The computational model of TensorFlow is based on a computational graph that can be executed on both the CPU and the GPU. [4] DNNRegressor, primitive of TensorFlow, was used as a basis to construct a required neural network.

Using TensorFlow, a three-layered neural network was created. An input layer consists of 50 formal neurons with a ReLu activation function:

$$f(x) = \begin{cases} 0, x < 0; \\ x, x \geq 1. \end{cases}$$

A hidden layer consists of 10 formal neurons with the ReLu activation function. An output layer consists of 1 formal neuron with a logistic activation function:

$$f(x) = \frac{1}{1 + e^{-x}}$$

To train the neural network, a dataset was built. Using the precise model computational experiments were performed with 34 different reactor configurations (burn-out of fuel, positions of the control system) and for each configuration, the criticality margin was obtained. From the obtained dataset, a training dataset (25 configurations) and a test dataset (9 configurations) were formed. For the final validation, all 34 configurations were used.

50,000 training epochs were performed for training artificial neural network. The training was made by the method of back propagation of the mean square error on the training dataset. Every 100 epochs, the mean square error was estimated on the test dataset. During the training process, the error was converging to zero without divergence.

Let's consider the results of validating the training of the artificial neural network: the average absolute approximation error is 0.0405; the maximum absolute approximation error is 0.1029; the average relative approximation error is 1.21%; the maximum relative approximation error is 3.13%.

The average time to calculate the criticality margin using an artificial neural network is 100 ms. The fig. 1 shows the calculated and approximated criticality margin.

Based on the results of the experiment, we can conclude: the construction of the approximation of the precision model was successful.

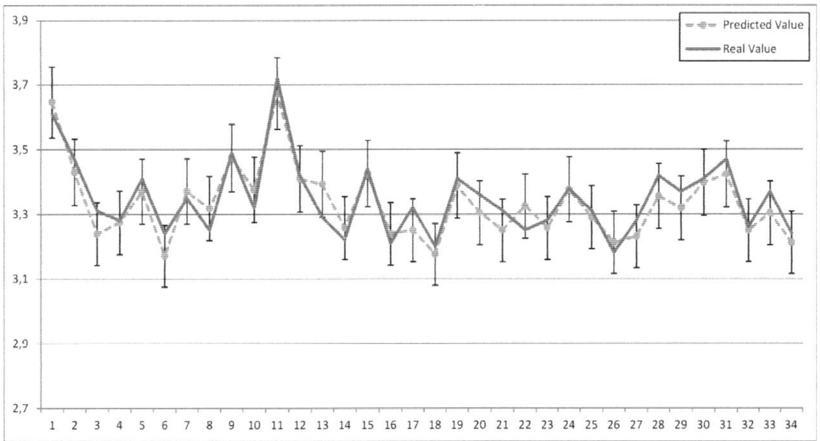

Figure 1. The calculated and approximated criticality margin charts.

3. Experiment #2: Measured Data Approximation

From the description of the experiment #1 it is clear that for the training of the artificial neural network only the data on the burning out of fuel assemblies, the position of the control systems, as well as data on the criticality margin of the reactor were used. Such data can be obtained not only from the precision model, but from the direct measurements that are made at the beginning of each campaign. Therefore, it is possible to construct an approximation based on the measured data.

For experiment #2, an artificial neural network was created using the TensorFlow library. The architecture of the neural network is completely identical to the network architecture described in experiment #1.

As a data set for training, the measured data of 24 reactor campaigns were taken. The data set was divided into 2 data sets: training data set (18 campaigns) and test data set (6 campaigns). To validate the trained neural network, data from all 24 campaigns were used.

50,000 training epochs were performed for training artificial neural network. The training was made by the method of back propagation of the mean square error on the training dataset. Every 100 epochs, the mean square error was estimated on the test dataset. During the training process, the error was converging to zero without divergence.

Let's consider the results of validating the training of the artificial neural network: the average absolute approximation error is 0.0412; the maximum absolute approximation error is 0.1159; the average relative approximation error is 1.26%; the maximum relative approximation error is 3.56%.

The average time to calculate the criticality margin using an artificial neural network is 100 ms. The fig. 2 shows the measured and approximated criticality margin.

Based on the results of the experiment, we can conclude: the construction of the approximation of the measured data was successful.

4. Conclusions and prospects

In this work, two artificial neural networks were created and trained to approximate the criticality margin of the WWR-c reactor (based on model and measured data). The resulting neural networks realize the correct approximation, have high accuracy, and also high speed of operation. The obtained artificial neural networks can be used to accelerate the preliminary calculations of the state of the reactor.

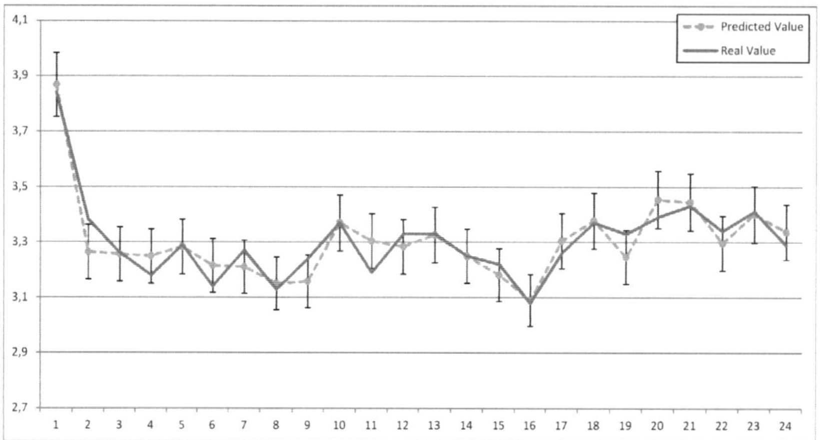

Figure 2. The measured and approximated criticality margin charts

In the future, the transfer of trained neural networks to the Karpov Institute of Physical Chemistry for use, improvement of the quality of approximation due to new data for training, application of the neural network approach for evaluating other complex processes of nuclear installations, development of an artificial intellectual helper for reactor technological processes.

References

[1] Kochnov O, Lukin N and Averin L 2008 *Nuclear and Radiation Safety* 18–25 ISSN 2218-8665
[2] Kolesov V, O Kochnov O, Volkov Y, Ukraintsev V and Fomin F 2011 *Известия вузов. Ядерная энергетика* 129–133
[3] Gorban' A 1998 *Siberian Journal of Numerical Mathematics* **1** 12–24
[4] Abadi M, Agarwal A, Barham P, Brevdo E, Chen Z, Citro C, Corrado G S, Davis A, Dean J, Devin M, Ghemawat S, Goodfellow I, Harp A, Irving G, Isard M, Jia Y, Jozefowicz R, Kaiser L, Kudlur M, Levenberg J, Mané D, Monga R, Moore S, Murray D, Olah C, Schuster M, Shlens J, Steiner B, Sutskever I, Talwar K, Tucker P, Vanhoucke V, Vasudevan V, Viégas F, Vinyals O, Warden P, Wattenberg M, Wicke M, Yu Y and Zheng X 2015 TensorFlow: Large-scale machine learning on heterogeneous systems software available from tensorflow.org URL `http://tensorflow.org/`

IOP Conf. Series: Journal of Physics: Conf. Series **945** (2017) 012032 doi:10.1088/1742-6596/945/1/012032

Algorithm for optimization of Viola–Jones object detection framework parameters

A D Egorov

National Research Nuclear University "MEPhI" (Moscow Engineering Physics Institute), Moscow

egorovalexeyd@gmail.com

Abstract. The Viola-Jones algorithm is one of the most popular algorithms for objects recognition in an image. This research paper deals with the possibilities of parametric optimization of the Viola-Jones algorithm to achieve maximum efficiency of the algorithm in specific environmental conditions. It is shown that with the use of additional modifications it is possible to increase the speed of the algorithm in a particular image by 2-5 times with the loss of accuracy and completeness of work by not more than 3-5%.

1. Introduction

Image recognition when analyzing a video stream or individual images is called computer vision. Computer vision is used in many modern technological solutions to ensure safety of objects and road traffic, personality analysis and recognition, building augmented reality. The mechanism for recognition a specific object in an image and video stream, for example, face, is an important element in the work of computer vision algorithms.

Recognition of a person's face in a complex task, including high optical interference and noise, is one of the main tasks to be solved when implementing remote technologies for monitoring their current functional and psychoemotional state [1, 2]. These technologies make it possible to carry out constant monitoring of the adequacy of behavior, first of all, of operational personnel managing hazardous objects to reduce the risk of technogenic accidents and catastrophes caused by the so-called human factor [3]. The developed algorithms and software for recognizing a person's face are subject to increased reliability and speed requirements, due to the need to process large video streams of data in real time [4]. The quality of the solution of this problem determines among other things the reliability of determining human bioparameters, related, for example, to the movement of the eyes, the level of microshivering of the facial areas, their temperature, and the reaction time [5].

2. About the Viola-Jones algorithm

One of the most common methods for object detection in an image is the Viola-Jones object detection framework. The Viola-Jones object detection framework [6, 7] was proposed by P. Viola and M. Jones in 2001. This method is based on the representation of the image in an integral form and the construction of a cascade of "weak" classifiers, which are called Haar-like features [8].

The Viola-Jones object detection framework consists in practice of the following steps:

1) Training a cascade classifier using an AdaBoost algorithm.
2) Preprocessing – constructing an integral image to optimize the work of the method.

3) Searching for an object in the image using Haar-like features.

The following parameters exist in the practical implementation of the algorithm:

1) Used cascade for objects search (trained before using the algorithm) – cas.
2) The change in the window size in the transition from one iteration of running the algorithm to another – sf. Min value of this parameter is 1,05. Maximum value is carried on maximum image size. The step of the parameter change is 0,05.
3) The number of objects recognized next to each other, on the basis of which it can be concluded that there is a sought object in this area – minn. Min value of this parameter is 1. Maximum is 10. The step of the parameter change is 1.
4) The initial size of the search box – mins. Min value of this parameter is carried on cascade min image size. In our case it is 20x20 pixels. Maximum value is carried on maximum image size. The step of the parameter change is 10.

Most of the papers in the study of the algorithm are devoted to improving the quality of cascade learning or the accuracy and speed of the comparison algorithm of the intended object. This research paper deals with the parameters of a particular implementation of the algorithm, as well as methods for optimizing them in order to obtain the highest accuracy with the smallest average operating time. Empirical comparison of various parameters of the method is given in article [9].

Optimization in the algorithm is based on the method of enumeration of the above parameters.

3. Mechanism of algorithm parameters selection

An image database (test set) with the desired objects under the conditions in question is necessary to select the parameters of the Viola-Jones algorithm for specific conditions. The set of intended objects (if they are present in the image) with the coordinates in this image should be put in line to each image in this database.

The following parameters calculated on the application of the method with a specific set of parameters to the database are to be introduced to evaluate the effectiveness of the method:

1) Accuracy – P – the ratio of the number of correctly found objects to the total number of found objects.
2) Completeness – R – the ratio of the number of correctly found objects to the total number of objects in the test set.
3) Based on the accuracy and completeness, it is proposed to use the aggregate parameter – F-measure (1).

$$F = \frac{2PR}{P+R} \tag{1}$$

4) It is proposed to introduce criteria (2,3,4) to work time tracking (T – the average time the method works on the test set)

$$Q_1 = a \cdot (1-P) + bT \tag{2}$$

$$Q_2 = a \cdot (1-R) + bT \tag{3}$$

$$Q_3 = a \cdot (1-F) + bT \tag{4}$$

5) It is proposed to use the parameter values for the considered criteria $a = 1, b = 1/\max(T)$.

4. Analysis of the results

The analysis of the results is carried out on the basis of 808 images with 1202 objects marked. Each image size is 400 x 400 pixels. Accuracy

The criterion Q_1 uses accuracy of work for calculation. Table 1 shows top-8 results obtained.

4.1. Completeness

The criterion Q_1 uses completeness of work for calculation. Table 2 shows top-8 results obtained.

Table 1. Selection of parameters for the criterion Q_1

cas	sf	minn	mins	true pos	false pos	false neg	time	P	R	Q_1
lbp_improved	1.4	5	70	417	0	785	2	1.000	0.347	0.012
lbp_improved	1.4	3	70	616	1	586	2	0.998	0.512	0.013
lbp	1.3	5	70	575	0	627	3	1.000	0.478	0.018
lbp	1.3	3	70	709	1	493	3	0.999	0.590	0.019
lbp	1.3	3	60	819	1	383	5	0.999	0.681	0.030
lbp	1.3	5	60	681	0	521	6	1.000	0.567	0.035
lbp	1.2	5	70	793	4	409	6	0.995	0.660	0.040
lbp_improved	1.4	3	40	714	3	488	7	0.996	0.594	0.045

Table 2. Selection of parameters for the criterion Q_2.

cas	sf	minn	mins	true pos	false pos	false neg	time	P	R	Q_2
lbp	1.1	1	70	1061	196	141	9	0.844	0.883	0.175
alt_improved	1.4	1	60	1066	30	136	10	0.973	0.887	0.177
lbp	1.1	1	60	1096	296	106	15	0.787	0.912	0.184
alt	1.2	1	70	1116	52	86	18	0.955	0.928	0.187
lbp	1.2	1	70	1024	107	178	6	0.905	0.852	0.187
alt2	1.2	1	70	1125	82	77	21	0.932	0.936	0.199
alt	1.2	1	60	1142	71	60	26	0.941	0.950	0.217
alt_improved	1.4	1	70	1010	10	192	8	0.990	0.840	0.211

4.2. F-measure

The criterion Q_3 uses F-measure of work for calculation. Table 3 shows top-8 results obtained.

Table 3. Selection of parameters for the criterion Q_3.

cas	sf	minn	mins	true pos	false pos	false neg	time	P	R	Q_2	F	Q_3
alt	1.1	3	40	1165	57	37	97	0.953	0.969	0.640	0.961	0.102
alt	1.1	3	40	1165	58	37	91	0.953	0.969	0.606	0.961	0.102
alt	1.1	3	30	1166	73	36	156	0.941	0.970	0.971	0.955	0.106
alt	1.1	3	30	1166	74	36	151	0.940	0.970	0.943	0.955	0.106
alt2	1.1	3	40	1166	87	36	109	0.931	0.970	0.706	0.950	0.112
alt2	1.1	3	40	1166	88	36	104	0.930	0.970	0.677	0.950	0.112
alt2	1.1	3	30	1168	111	34	177	0.913	0.972	1.085	0.942	0.116
alt2	1.1	3	30	1168	112	34	171	0.913	0.972	1.051	0.941	0.117

5. Conclusions and recommendations

On the basis of the results, inference should be drawn that it is not expedient to use criterion of selection of parameters which is estimated only on accuracy of algorithm work, as the completeness does not exceed 31.8% for 8 best results.

It is necessary to increase the weight of completeness when calculating the criterion. For example, we take the coefficient a for the criterion Q_2 equal to three and consider the results (table 4).

It is seen that the highest completeness is 0.963 with the accuracy of 0.894, taking into account the error of the criterion being calculated. This accuracy is less by 5.9% and the completeness is less than 0.6% than the optimal result, calculated using the criterionQ_3. However, the time the method works when taking parameters from Q_2 differs in 2.2 times from the time the method works when the parameters are taken from Q_3. The greatest gain in time – 5.38 times – is achieved with the difference of the accuracy of 0.2% and the completeness of 4.1%.

Table 4. Selection of parameters for the criterion Q_2 with coefficient 3.

cas	sf	minn	mins	true pos	false pos	false neg	time	P	R	Q_2
alt	1.2	1	60	1142	71	60	26	0.941	0.950	0.297
alt2	1.2	1	70	1125	82	77	21	0.932	0.936	0.311
alt	1.2	1	70	1116	52	86	18	0.955	0.928	0.316
alt2	1.2	1	60	1144	124	58	31	0.902	0.952	0.320
alt	1.1	1	70	1131	95	71	26	0.923	0.941	0.324
alt	1.2	1	50	1156	95	46	39	0.924	0.962	0.335
alt	1.1	1	60	1158	138	44	40	0.894	0.963	0.336
alt2	1.2	1	60	1144	123	58	36	0.903	0.952	0.348

It is recommended to introduce an additional check for the presence of objects on the found object to increase the accuracy of the work. For example, for a face it is the search for the eyes, mouth or nose on the supposed face found.

Thus, the algorithm for selecting parameters consists of the following steps:

1) Building a database of objects for the current environmental conditions.
2) Selection of parameters based on the criterion Q_2 or Q_3 depending on the business requirements of the system.
3) Increase the accuracy of work due to the modification of the additional search used.

Laboratory tests of developed software tools for optimizing human face recognition algorithms have confirmed their high efficiency.

The application of the developed algorithm for selecting parameters for face recognition in software and hardware complexes for remote monitoring of the current functional and psychoemotional state of a person will ensure continuous reliable registration of its bioparameters using optical technologies.

Acknowledgement
The research is carried out within the framework of the theme 00-2-026-4209 "Approbation of the technology for reducing the risk of occurrence and mitigation of the consequences of man-induced disasters by minimizing the influence of the human factor on the reliability and trouble-free operation of NPPs and other hazardous objects" in accordance with the competitive part of the MEPhI government contract for 2017-2019.

References

[1] Aliushin M V and Kolobashkina L V 2014 Psychology questions *Monitoring of bioparameters of the person on the basis of remote technologies* №6 pp. 135–144.

[2] Aliushin M V, Aliushin A V, Andriushina L O, Kolobashkina L V and Pshenin V V 2013 Global nuclear safety *Remote and not contact technologies of registration of bioparameters of operation personnel as control facility human factor and increases in the NPP safety.*, №3(8) pp. 69–77

[3] Alyushin M V, Alyushin V M, Kolobashkina L V 2016 Psychology issues *Methodological aspects of the automated forecasting of emergency situations of technogenic origin* №2, pp. 83– 90

[4] Aliushin M V, Aliushin AV, Belopolskii V M, Kolobashkina L V and Ushakov V D 2013 Global nuclear safety *Optical technologies for systems of monitoring of the current functional state of quick structure of management of nuclear power objects* №2(7), pp. 69–

77

[5] Alyushin M V and Alyushin V M 2015 Psychology issues *Technique of measurement of time of operator reaction*, №5, pp. 157–165

[6] Viola P. and Jones M. 2001 proceedings IEEE Conf. on Computer Vision and Pattern Recognition (CVPR 2001) *Rapid Object Detection using a Boosted Cascade of Simple Features*

[7] Viola P. and Jones M. 2004 International Journal of Computer Vision *Robust real-time face detection*, vol. 57, no. 2., pp.137–154

[8] Whitehill J. and Omlin C.W. 2006 Automatic Face and Gesture Recognition *Haar features for FACS AU recognition* pp. 101-106.

[9] Egorov A D, Shtanko A N and Minin P E 2015 Bulletin of the Lebedev Physics Institute *Selection of Viola-Jones Algorithm Parameters for Specific Conditions* **42** No 8 pp 244-248.

Perspectives of intellectual processing of large volumes of astronomical data using neural networks

A A Gorbunov[1], **E A Isaev**[1,2], **V A Samodurov**[1,2]

[1]National Research University Higher School of Economics
[2]Pushchino Radio Astronomy Observatory of Astro Space Center of P.N. Lebedev Physics Institute, RAS, Pushino, 142290

agorbunov@hse.ru, eisaev@hse.ru, vsamodurov@hse.ru

Abstract. In the process of astronomical observations vast amounts of data are collected. BSA (Big Scanning Antenna) LPI used in the study of impulse phenomena, daily logs 87.5 GB of data (32 TB per year). This data has important implications for both short-and long-term monitoring of various classes of radio sources (including radio transients of different nature), monitoring the Earth's ionosphere, the interplanetary and the interstellar plasma, the search and monitoring of different classes of radio sources. In the framework of the studies discovered 83096 individual pulse events (in the interval of the study highlighted July 2012 - October 2013), which may correspond to pulsars, twinkling springs, and a rapid radio transients. Detected impulse events are supposed to be used to filter subsequent observations. The study suggests approach, using the creation of the multilayered artificial neural network, which processes the input raw data and after processing, by the hidden layer, the output layer produces a class of impulsive phenomena.

1. Formulation of the problem

The basis for the research in this study is the data recorded by Big Scanning Antenna (BSA) LPI. BSA LPI daily registers 87.5 GB of data (32 TB per year). For structuring the flow of "big data" researchers use a technique based on the methods of mathematical statistics [1,2]. In some cases, automatic classification applied for statistical processing produces incorrect results and the resulting impulses have to be checked and sorted visually [1]. In this regard, there is a need for more effective classification methods, allowing to obtain more reliable results.

Among the papers considering the problems of classification of impulse phenomena are noticeable developments using a neural network. For example, in [3] shows the results of using multilayer perceptron and self-organizing Kohonen network to solve the problem of recognition of the impulses of the meteors. There is also a patented method that allows determining the structure and demodulation of the signal with unknown structure by using complex multi-stage study characteristics [4]. At a certain stage of this method supposes the use of Kohonen neural network [4].

It should be noted that the main focus of studies aimed at determining the possibilities of using neural networks to classify impulse phenomena. With the current state of research, it remains unclear whether it is possible to use classifiers based on neural networks to process large amounts of astronomical data. In the work [1] were identified 83096 impulse phenomena. It is required to determine the possibility of a multilayer neural network for classification the raw data. Conducting practical experiments for training artificial neural networks with thousands of elements shows that it takes several tens of hours to the network began to show signs of intelligent behavior [5].

The aim of this work is to develop the neural network for structuring and processing large amounts of astronomical data

To achieve the goal the following tasks are set:

- Analysis of the structure of the input/output data necessary for the design of a neural network;
- Conducting experiments using different types of neural networks.

2. Analysis of the structure of a stream of input/output data

In order to determine the number of neurons in the input layer of the neural network, it is necessary to explore the structure of the data stream. Astronomical data [1] processed by two recorders, consisting of 6 modules. Each module registers 8 signals (rays). The signal for each beam is divided into 6 bands of frequency ranges. Additionally, it registers a signal with a common bandwidth of the radio telescope. The calculation of the statistical characteristics of the input data for each band (from 1 to 6), for each of the 96 beams is produced every 5 seconds.

Thus, to classify the 5-second observations requires a neural network with a number of inputs 96*6 (number of rays*number of bands) or 576.

The number of neurons in the output layer is determined by the number of observed classes of impulse phenomena. The result of statistical processing of data, registered by BSA LPI during the period from 2012 to 2013, has been allocated the following classes of phenomena:

- impulses of technogenic character of various kinds;
- hardware failures data;
- impulses having the character of single flickering radio sources of different types;
- impulses having well-defined space character.

Thus, the output layer consists of 4 neurons, the outputs of which will indicate a particular class of observable data.

3. Conducting experiments using different types of neural networks

Experiments on the recognition of different classes of impulse phenomena were carried out for a multilayer perceptron and self-organizing Kohonen network. Neural network data types were used in [3] to identify observations corresponding to the meteors. For training neural networks used 5 –minute recording at a frequency of 2.5 kHz. The result of this research is a neural network, is able to divide the observations into 2 classes: the meteors and non- meteors.

The basis of the experiments, the results of which are given in the framework of this work are the classes of impulse phenomena identified during observations in the period from 2012 to 2013. The complex flow of astronomical data (see section 2) must be classified by the neural network. Statistical processing allowed to reveal 83096 different phenomena, which can be divided into 4 classes (see section. 2). This sample is used as training dataset. It is expected that after training, the neural network will be able to divide the observations into 4 classes.

Implementation and experiments with neural networks have been performed using MATLAB tools.

3.1 Multilayer perceptron

A multilayer perceptron [5] consists of input, hidden, output layer. The hidden layer may consist of one or more layers. The input layer has no synaptic weights, and thus not trained. A multilayer perceptron is trained with the teacher, that is, the input is a vector of observations, and in the output, and layer checks the consistency of the result of the neural network by the previously known answer.

If the expected response does not match the output result, the adjustment of synaptic weights of the hidden layer is produced. Thus, using the examples you can get from a neural network solving the task: the classification of astronomical observations. The neural network is trained using learning algorithms: backpropagation, gradient descent, Levenberg –Marquardt.

The results of applying different learning algorithms are presented in table 1.

Table 1. The results of training the neural network

	Training sample size	Test sample size	Recognition rate (training sample), %	Recognition rate (test sample), %
Backward propagation			84,69%	69,43%
Gradient descent	62322	20774	93,41%	59,64%
Levenberg–Marquardt algorithm			98,73%	88,97%

Pre-fetch from 83096 classified observations, divided into training (75%) and test sets (25%). The best classification result of test data achieved when using the training algorithm of Levenberg – Marquardt. Percentage of recognition data to the training sample is not increased 100% in order to avoid overfitting. The duration of the learning process was 1000 epochs.

3.2 Kohonen network

In contrast to the multilayer perceptron that uses the algorithms of supervised learning, Kohonen network [5,6] does not require the filing of the expected values of the classification at the output layer. The Kohonen network consists of an input and topological layers.

As a result of self-organization of a network of neurons in the topological layer are grouped together, forming clusters. Thus, when the input of the unclassified observations-organized neural network determines its membership to one of the clusters. In our case, for self-organizing network, observations from 4 classes of impulse phenomena forming the training sample is sequentially applied to the input Kohonen neural network and thereby cause the neurons to be grouped into clusters

4. Evaluation of results

After training the neural network (p.3.1, 3.2) were used for the classification of astronomical observations. The classification results are shown in tables 2, 3.

Table 2. The results of the classification using the Kohonen network

	Recognized observations	Unrecognized observations	Recognized as the objects of another class
impulses of technogenic character of various kinds (3116)	2939	177	14
	94,31%	**5,69%**	**8%**
hardware failures data (4155)	3871	283	5
	93,18%	**6,82%**	**2%**
impulses having the character of single flickering radio sources of different types (5194)	4957	237	50
	95,44%	**4,56%**	**21%**
impulses having well-defined space character (8310)	8105	204	26
	97,54%	**2,46%**	**13%**

Using the methods of classification, based on methods of mathematical statistics [1], misclassified remain 10-15 % of cases.

Table 3. The results of classification using multilayer perceptron

	Recognized observations	Unrecognized observations	Recognized as the objects of another class
impulses of technogenic character of various kinds (3116)	2732	385	57
	87,66%	**12,34%**	**15%**
hardware failures data (4155)	3639	516	30
	87,58%	**12,42%**	**6%**
impulses having the character of single flickering radio sources of different types (5194)	4656	537	63
	89,66%	**10,34%**	**12%**
impulses having well-defined space character (8310)	7558	751	47
	90,96%	**9,04%**	**6%**

The use of Kohonen neural network shows the smallest percentage of unrecognized data (average 5%) as compared to the use of the classifier based on multilayer perceptron (average 11%) and in comparison with the methods of statistical processing [1].

5. Conclusion

In this work, we investigate the question of the applicability of different technologies of neural networks for classification of large volumes of astronomical observations. As the basis of research used data from round-the-clock observation was BSA LPI in the period 2012 – 2013 the analysis of the structure of the flow of astronomical data. Based on the results of the research presented in [1] determined the resulting classes is required to recognize using neural network technology.

Experiments on learning the two types of neural networks: multilayer perceptron and self-organizing Kohonen network. The classification accuracy of astronomical observations made with neural networks is higher than when using the methods of [1], based on the methods of mathematical statistics.

References

[1] Samodurov V A, Dumsky D V, Isaev E, Rodin A E, Kazancev A N, Fedorova V A, Belyatskij Y A 2016 The daily 110 Mhz radio wave sky survey: statistical analysis of impulse phenomena from observation in 2012-2013 *Odessa Astronomical Publications* **29** pp 167-170.

[2] Taylor G B et al: 2012 First Light for the First Station of the Long Wavelength Array *J. Astron. Instrum.* **1** pp 1–56

[3] Victor S R, Ca˘ta˘lin B 2015 Automatic Analysis of Radio Meteor Events Using Neural Networks *J Earth Moon Planets*. pp 101-113

[4] Kuzovnikov A V, Anzhina V A, Pashkov A E et al 2010 Sposob opredeleniya struktury i demodulyatsii signala s neizvestnoy strukturoy *Patent RF № 2386165 / 10*

[5] Tadeusevich R I et al 2011 Elementarnoe vvedenie v tekhnologiyu neyronnykh setey *Goryachaya liniya telekom*

[6] Kohonen T 1990 The self-organizing map. *Proc. IEEE* **78** pp 1464–1480

[7] Konovalenko A A et al 2016 Astronomicheskie issledovaniya s pomoshch'yu malorazmernykh nizkochastotnykh radioteleskopov novogo pokoleniya *Radiofizika i radioastronomiya* **21** pp 83-131.

[8] Tyul'bashev S A, Tyul'bashev V S 2015 Otkrytie novykh pul'sarov na radioteleskope BSA FIAN *Astronomicheskiy tsirkulyar* 1624

[9] Rodin A E, Samodurov V A, Oreshko V V 2015 Bystryy metod obnaruzheniya periodicheskikh radioistochnikov na BSA FIAN *Astronomicheskiy tsirkulyar* 1629